The California Earthquake of 1906

Edited by David Starr Jordan

Articles by

David Starr Jordan, President Stanford University

John Casper Branner, Vice-President and Professor of Geology at Stanford University

Charles Derleth, Jr., Associate Professor of Structural Engineering, University of California

Grove Karl Gilbert, United States Geological Survey

Stephen Taber, Stanford University

F. Omori, Sc. D., Member of the Imperial Earthquake Investigation Committee, Tokyo, Japan

Harold W. Fairbanks, Berkeley, Cal.

Mary Austin, Author of *Land of Little Rain*, *The Flock*, Etc.

Originally Published by A.M. Robertson,
San Francisco, California, 1907.

Reformatted and Published by Jessica Duke,
Las Vegas, Nevada, 2011.

Book Layout and Cover Design by Jessica Duke.

Printed in the United States of America.
ISBN: 978-1-61393-008-3

The original publication with which this reprint was based was very faded and showed deterioration from age. The images have been reproduced with the best quality possible. The publisher believes that the ability to view these images that captured this historic tragedy, regardless of their quality, is, nonetheless, priceless.

Prefatory Note

After the Great Earthquake of 1906, a number of accounts of the matters of geology and engineering involved, appeared in the current magazines. One of these, published by the editor of the present volume, appeared in the Popular Science Monthly. The present publisher, Mr. Robertson, asked the privilege of reprinting this article, with such others of similar nature, as might be available for the purpose of a volume treating of the scientific aspects of the Great Earthquake.

The essays thus chosen constitute the present volume. Those of Professor Branner and Mrs. Austin are reprinted from Out West, by the courtesy of the editors, Charles F. Lummis and Charles Amadon Moody; the essay of Mr. Gilbert and that of the present writer are reprinted from the Popular Science Monthly by the courtesy of the editor, Professor J. McKean Cattell; the essay of Mr. Taber is reprinted from The Journal of Geology by the courtesy of Professor Thomas Chrowder Chamberlain; that of Professor Omori, from the Bulletin of the Imperial Earthquake Investigation Committee of Japan, by the courtesy of this Commission; that of Mr. Fairbanks, from the Bulletin of the California Physical Geography Club, by the courtesy of the Club. Professor Derleth's essay was written expressly for the purpose of this volume, appearing here for the first time. It is believed that the series of articles give a clear, comprehensive, and accurate view of the Great Earthquake and its associated phenomena.

DAVID STARR JORDAN.

Stanford University,

April 18, 1907.

CONTENTS.

The Town Columns.

The Earthquake Rift of April, 1906

By David Starr Jordan,

President Stanford University

There are two sets of disturbances which shake the crust of the earth and therefore go by the name of earthquakes. Eruptive earthquakes are explosions, usually of steam, about a volcano. Tectonic earthquakes are breaks in the overloaded or overstrained crust of the earth, and, for the most part, have nothing to do with the steam vents we call volcanoes. To the last class most earthquakes belong, certainly almost all that have been felt within the United States.

Again, under the name of earthquake we include two very different sets of phenomena, the one the rock-rift or fault, which is the disturbance itself, the other the spreading or interfering waves set in motion by the parting, shearing and grinding of the sundered walls of rocks in the earthquake fault. It is the jarring waves extending in widening and interfering circles which do the mischief to man and his affairs. It is the rift of rock which sends these waves forth on their blind mission of confusion and destruction.

In every tectonic earthquake there is somewhere a fault or rift of rock, with some sort of displacement, permanent or temporary, of the relations of the two sides. In extreme cases, this break extends for miles in a straight line, breaking the surface soil and passing downward to a depth which can be only guessed at, five or ten miles perhaps, probably as far down as the crust is rigid. There are undoubtedly destructive earthquakes in which the soil is not broken over the rift of rock, but as a rule, in violent disturbances, the crack comes to the surface, breaking through the overlying soil. In all severe earthquakes there are, moreover, breaks or fissures in the earth having no connection with the fault itself. These are slumps or landslides, and geologically they signify but little. They mean simply that loose soil has been shaken down. They do not go down into the underlying rock. From the true earthquake crack they may usually be known at once, because their course is determined by the topography. They are not straight. The true earthquake rift moves on in straight lines, broadly speaking, careless of topography. But topography is not careless of the earthquake rift. On either side of it, for perhaps hundreds of feet, the rocks are crushed to flinders by the impact and grinding of the opposed walls. An old fault is therefore marked by an excess of erosion. A valley or saddle marks its general course. Streams choose it for their basins, and when it crosses a mountain the softened rock yields to form a saddle or other form of depression. For these reasons, an earthquake fault is often marked in California by successions of dairies and of reservoirs. The valleys thus formed are fertile and well watered. For the most part, in much-faulted regions, such as form the rim of the Pacific, each earthquake rift follows the line of an old fault, and the original break goes back to the mountain-making periods of Tertiary times. The California earthquake of 1906 follows the axis of a very ancient break, the 'Portolá-Tomales fault,' also called the `San Andreas fault,' first studied, so far as I know, by Dr. John C. Branner in 1891. In this fault hundreds of thousands of earthquakes, large and small, have preceded the recent one. In it the aggregate displacement horizontally has been very great, and the aggregate vertical displacement produced by all its many earthquakes, as shown by the rock strata on either side of it, exceeds half a mile.

From the rift at times in the past, masses of molten rock have flowed out. Of such origin is the cliff of basaltic columns near San Francisquito Creek, on the Stanford University campus. Much more recent flows of black lava occur to the southwest of Stanford University and numerous dikes of lava occur for the whole length of the Santa Clara Valley; these have not flowed from volcanoes but in times long past have escaped from rifts in the rock-producing earthquakes.

An overflow of lava of this kind seems to be the origin of the picturesque 'Marysville Buttes' in Sutter County, and of Anacapa Island in the Santa Barbara channel.

It is the purpose of this article to trace the earthquake rift of April 18, 1906, across the map of California.

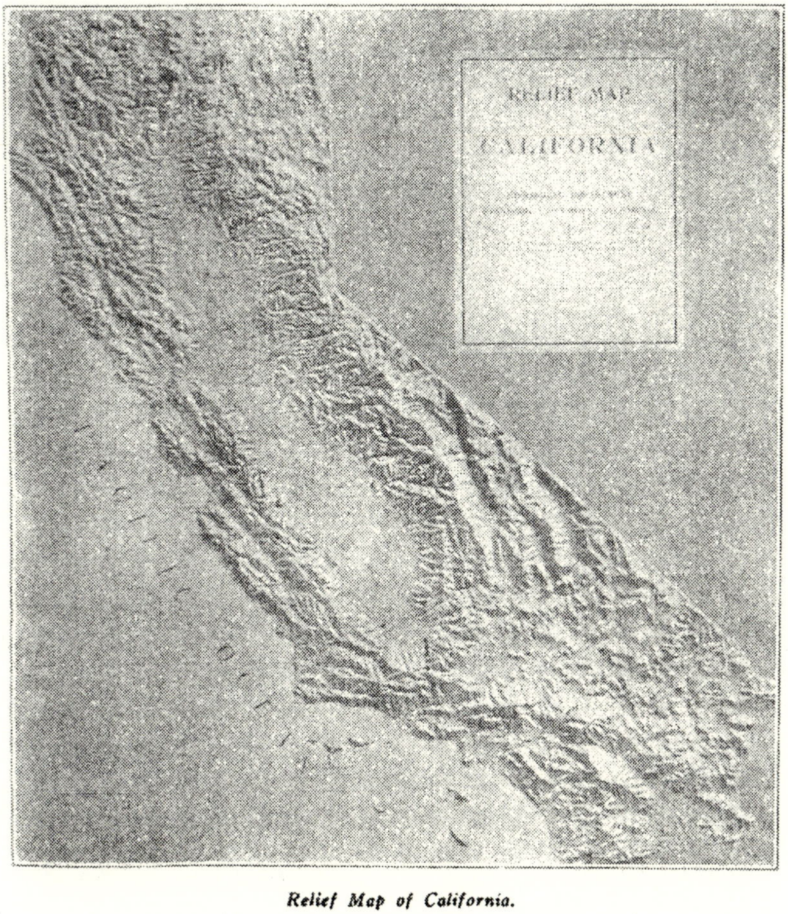

Relief Map of California.

Earthquake Rift, As It Comes up from the Sea at Point Arena, Mendocino County.

The accompanying photograph of a relief map by Dr. Noah Fields Drake will show the topography of the state. In California there are multitudes of valleys of various kinds. Those formed by water and ice surface erosion are variously curved and ramified. Such are the mountain cañons of the west flanks of the Sierras. Those valleys formed or marked by earthquake cracks have almost invariably straight axes. These extend in general toward the north-northwest, more or less distinctly parallel with each other, and often intersected by cross-faults.

Examples of faulted valleys are the great valley of the Sacramento and San Joaquin, the Santa Clara Valley, San Francisco Bay, with the Valley of Santa Rosa, Eel River Valley, the Santa Catalina Channel, Owens River, the San Jacinto Valley and many others. A cross-fault extends from Monterey Bay up the valley of the Pájaro River. In some of these faults earthquakes have taken place in historic times, in others no break has been noted save that recorded in the rocks. Dr. Branner has compared a fault to a break in a bone. It represents a weak place which will give in a time of strain. On the other hand, if not freshly broken, it will tend with time to heal. A broken bone will be naturally renewed. A faulted rock bed will be cemented in the course of ages of pressure and of cementation. Some of these rifts have been cemented and closed by their own lava-flows. This seems to have occurred in the case of the two ridges which bound the valley of Napa.

Fissure and Landslip, San Jacinto Valley, 1897.

The most interesting of these breaks in California is that recorded as the Portolá-Tomales fault. Its course can be plainly traced on the relief map. It enters the shore from the sea near the mouth of Alder Creek, to the north of the low headland called Point Arena, in Mendocino County on the north, and runs to Chittenden, on the Pájaro River, in Monterey County, on the south. The line is almost perfectly straight, and its course and direction can be determined by placing a ruler on the map, using the line of Tomales Bay as an axis.

This long, narrow, straight inlet is a resultant of past earthquakes, probably beginning in Tertiary times. It is bounded on the west by mountains which have their origin in some ancient upward thrust of the walls on the west side of the ancient fault. From Chittenden the same fault extends to the southward along the axis of the Gavilan Mountains for perhaps 300 miles more, quite straight, as far as Monte Pinos in Ventura County, past Priest Valley, Cholame and the Carisa desert. From Monte Pinos, it is said to curve to the eastward past San Bernardino and San Jacinto to near Yuma, but as to this eastward extension the present writer has no exact knowledge.

Alder Creek Bridge, Mendocino County. The Earthquake Rift Is near the Middle of the Picture.

On the eighteenth of April the trouble began in the sea. Just where, we may find out later. We know that the center is in the sea because where the rift enters the land the motion was more violent and the effects of the shock greater than at any other point along its extent. As the opened rift can be traced for 192 miles across the land to the southward from Point Arena, it is safe to say that it goes as far to the northward under the sea. The steamer *Argo* crossing it the moment of the earthquake, off Mendocino, ninety miles to the northward of Point Arena, bears witness to this fact. The captain thought that he had struck a raft of logs, so fierce and hard were the shocks of the waves in the water. The movements were short, quick and violent, not forming a tidal wave, but a strange choppy sea. For the time being all rollers and surf were broken up. Off the bold headland of Cape Mendocino is a deep sub-marine valley, to the west of which is a high mountain which does not rise to the surface of the ocean. In the channel between the cape and the submerged mountain the earthquake rift may be supposed to run. In this channel numerous earthquake shocks have been recorded by different passing vessels. If not itself a center of disturbance, it records the line along which great disturbances have frequently passed.

Tomales, Marin County. The North Shore Railroad and the Earthquake Rift.

The rift struck the land at the mouth of Alder Creek, above Point Arena. It crept over the hill as a deep furrow in the black, sticky adobe, veering a little to left or right according to the resistance of the soil, but always keeping in a straight line in its general direction. It may be imagined as a sort of devouring dragon, leaving its trail on the hills and destroying the works of man wherever it passes. It is hard in following its course, not to think of it as endowed with a sort of wicked life. Its movement is properly from north to south, but the opening of the great fault seems to have been really instantaneous. It took place at 5:13 a. m. and the waves lasted forty-seven seconds. It may be noted in passing that the complication of the waves at any one point was mainly due to the great length of the rift. A point immediately near the crack felt mainly the first great shock, its wave and the return wave. A point farther away felt the wave and its return jolt, followed at once by waves from farther to the north and farther to the south, these waves becoming more and more opposed to one another. The waves would then augment, neutralize, override and otherwise modify one another, the final result being the violent twisting motion, the most remarkable trait of the latter portion of the earthquake in question.

Coming over the first ridge, from the sea, the rift passed under the long bridge over Alder Creek.

The land on the west side of the bridge was jerked sixteen feet to the north; or that on the east sixteen feet to the south—only a careful re-survey of the region can tell us which. For this information we must depend on the work of the U.S. Coast and Geodetic Survey. Or it may be that both sides went to the northward, but the west side pulled away, distancing the other by sixteen feet. In any case, the bridge was torn to splinters, and the crack went on, always the west side some sixteen and a half feet to the northward, though the sticky soil tends to lag back, and not every place shows the maximum of shearing or horizontal displacement. Passing under a barn, the rift tore it to splinters. The spreading wave displaced or destroyed most of the houses in the villages of Manchester and Point Arena, wrecking the magnificent

Point Arena. Picket Fence Was in One Continuous Line. Photograph Shows Short Section Put in to Fill up Offset by Land on the West Side.

lighthouse of solid masonry on the Point itself. In low ground the rift formed successions of little ponds. On hillsides the lower side of the crack fell away like a drivelling lower lip, leaving an open chasm, ten to twenty feet in apparent depth. On level hard ground the soil like the rock below closed with a snap a little tighter than it was before. Line fences were broken and sheered from sixteen to twenty feet. Lines of trees met with similar readjustments. In Mendocino County the horizontal displacement is about sixteen feet. In Marin County, wherever it is exactly measured, it is sixteen feet and seven inches. Southward it becomes less. In San Mateo County it is six to eight feet, and at the Pájaro Bridge at Chittenden, near which point the open fault ceases, the western pier was moved northward about eighteen inches. This shifting of position, evident along the line of the crack, seems to have included the whole region, mountains and valleys through which the crack passes. Either the region to the westward with the Santa Cruz Mountains and the mountains called Sobrante de la Punta de los Reyes have been stretched out toward the northward or else the region on the east side, including most of California, has been correspondingly

Landslip at Sobrante.

humped up. It is impossible at present to say which is the fact, perhaps both. The vertical displacement is small. To the north of San Francisco the west side has been raised two or three feet. To the southward the slight relative change in elevation—two or three feet—is in favor of the east side.

The rift left the pastures of Point Arena, passing up Gualala River, always in a straight line, making havoc among the redwood trees and still greater havoc in the town of Fort Bragg which was in part shaken down, and afterwards burned. From Mendocino County it passes into the sea, where it runs close along the coast of Sonoma County past Fort Ross, throwing down everything movable in this and other towns. It then crosses Bodega Head and again falls into the sea, where it passes up the axis of Tomales Bay. At the head of the bay its course through the tules or bulrushes looks like a swath through a grain field. Through this region (Marin County) the shock

Landslip at Sobrante.

was very violent, and numerous cracks parallel with the main crack in the bay extended along the shores. In the town of Tomales, much and varied mischief was done. The parallel cracks toyed with miles of the North Shore Railroad between Tomales and Point Reyes. At Marshall the humble hotel was thrown bodily—and upright—into the bay, the boarders unharmed; and at aristocratic Inverness, on Tomales Bay, three summer cottages suffered the same fate. A fisherman in the bay reports that the waters of Tomales Bay receded, leaving his boat in the mud.

Afterwards they came back in a 'great wave, which looked a hundred feet high, but which was probably not more than ten.'

At Point Reyes Station at the head of Tomales Bay the 5:15 train for San Francisco was just ready. The conductor had just swung himself on when the train gave a great lurch to the east, followed by another to the west, which threw the whole train on its side. The astonished conductor dropped off as it went over, and at sight of the falling chimneys and breaking windows of the station, he understood that it was the Temblor. The fireman turned to jump from the engine to the west when the return shock came. He then leaped to the east and borrowing a Kodak he took the picture of the train here presented.

Marshall Hotel, Thrown into Tomales Bay.

Earthquake Crack in Country Road from Olema to Point Reyes.

Earthquake Rift, Freeman's Ranch, near Tomales Bay.

Paper Mill Creek runs past the same village, a considerable stream, noteworthy lately from the experiments in stocking it with king salmon. The two banks of the stream were forced toward each other so that the length of the bridge was shortened by about six feet and the bridge was correspondingly humped at its north end, an arch about six feet high being forced up.

From Point Reyes Station (at the base of the large peninsula called Point Reyes) the earthquake rift passed along the Inverness Road to Olema, where all the houses not standing on rock foundation were thrown from three to six feet to the westward, toward the crack itself.

Skinner's Ranch is a large dairy near Olema. The house stands near the road, a dairy house some thirty feet to the south of it, and a large barn with cowyard just behind that. A row of large cypress

Earthquake Rift, Freeman's Ranch.

trees stood just before the house on the roadside, between them and the house a little rose garden, to the south of these, opposite and partly behind the dairy, a group or row of large eucalyptus trees. The earthquake rift passed directly in front of the house, between the buildings and the road. All that stood to the westward of the crack was violently jerked to the north a distance of sixteen feet seven inches, or it may be that the east side moved an equal distance to the south. If Mr. Skinner had chanced to look at the right instant he would have seen the whole row of cypress trees file past his window to take their station in front of the dairy, taking the rose garden with them. A few raspberry bushes came from farther north to take, partly, the place of the roses. The eucalyptus trees in front of the dairy moved on to a position opposite the barn, and one detached from the others and to the westward of the crack was left near the head of the line instead of at its foot. The crack passes obliquely under the barn, entering it at the northwest corner and leaving it at the middle of its posterior or southwest side. The barn remained intact, thanks to its weak foundation, for the east side pulled loose from the ground, and the barn went northward with

Train Overturned by Earthquake, Point Reyes Station, Marin County, Cal.

the west side. Sixteen and one half feet of its former foundation at the southeastern corner is exposed. A driveway under the barn is divided in the middle. You pass in on the east side, the western half is sixteen and one half feet to the north of the entrance and completely blocked in the middle. Under each of the east windows of the barn stood a pile of manure. Each pile is intact, sixteen and one half feet south of the window to which it belongs. The people at the ranch were milking at the time of the shock. Each man and cow was thrown to the ground and it took two hours to get the frightened cattle back into the `corral.' The stone steps to the basement of the dairy, on the east side of the crack, now stand sixteen feet seven inches to the southward of the door to which they led. About Skinner's, line fences and water pipes crossing the fault were broken, a break of sixteen and one half feet being left is each case, the west side of the fault in all cases more or less overriding the other.

Earthquake Rift, Olema.

In the matter of line fences interesting legal problems are raised. Were the farms on the west stretched sixteen and one half feet or those on the east side crowded together to the same amount? If either, who stands the loss and what store can be set on ancient landmarks? The observations of Dr. Grove K. Gilbert, verified by other geologists, leave little doubt that both sides of the rift moved, the west side to the northward, the east side to thesouthward. How far from the rift this motion extends to the east and to the west, is yet to be shown, manifestly it is a local, not a continental, change of position.

Next to the Skinner Ranch is the Shafter Ranch. Here the houses and barns are on the east side of the crack, but the transposition of roads, trees and fences was the same in kind. The rift passed through the corral, and one of the astonished cows dropped into it, soon falling deeply till only rump and tail were visible. The hysterical dogs barked at her, the water came into the rift, and the dairymen, doubtless with a sense of the impotence to struggle against fate, buried her in the grave from which they could not rescue her.

Skinner's Ranch, Olema.

Crossing the valley the rift split a small hill, throwing down four large spruce trees, all of which fell at right angles to the crack. A very large oak tree standing on level ground was shoved violently, still standing, sixteen and one half feet to the southward into the base of the riven hill, or perhaps the western half of the hill was shoved violently about the tree.

On through the valley of Olema went the rift, past more dairies, but leaving their buildings altogether to the east. Crossing the road above Bolinas, the two sides of the highway are rudely separated. Reaching Bolinas Bay, the rift is visible in the mud at low tide, and careful observers report the sea-bottom to the westward, along Duxbury Reef, to be raised two or three feet. The gatherers of abalone shells venture out into regions of sea-bottom formerly inaccessible at the lowest tides. On the east side of Bolinas Bay the clams are hopelessly buried. At Bolinas the pretty Flagstaff Inn was thrown bodily into the sea and completely wrecked. The crack again enters the sea, passing across the entrance to the Golden Gate five

Earthquake Rift, Morrill's Ranch, Skylands, Santa Cruz County.

or six miles west of the center of San Francisco, and giving to that breezy and joyous town a jolt which will live in history, and causing through the subsequent fire a greater destruction of the results of human effort than was ever known before in the records of the world. The rift reached the shore again at Mussel Rock to the southwest of San Francisco. Here the cliff was hurled down, a gradual incline was made a steep one and four thousand feet of newly graded railroad was thrown into the sea. It passed up the narrow valley of San Andreas, not harming the reservoir on account of the splendidly built dams of reenforced concrete, but wrecking all the water mains entering San Francisco from the great reservoirs, Crystal Springs, San Andreas and Pilarcitos. The dam of the Crystal Springs reservoir, across the fault line, was also so well built that the visible crack passed around it along the bank by its side, returning afterwards to its former direction. The bleak and boulder-strewn saddle called Cañada del Raymundo, scarred by previous

Earthquake Rift, Morrill's Orchard, Santa Cruz County.

earthquakes, was then passed, and beyond it the narrow, fertile valley of Portolá, named for its discoverer, Gaspar de Portolá, the first governor of California, the discoverer of San Francisco Bay. The crack runs along the base of the Sierra Morena, four to five miles west of Stanford University, at the corner of its generous campus, to the head of Portolá reservoir; then ascends in a cañon to a saddle on the summit, connecting two parallel ranges, Monte Bello to the east and Castle Rock to the west. Down from the saddle between these runs Stevens Creek (Arroyo de San José de Cupertino) and down this creek went the earthquake crack, tearing up the road behind it, and throwing down landslides from every steep slope. Stevens Creek is made up from the union of two streams which meet from opposite directions. The crack descends the one and remounts impartially in the valley of the other. Both streams follow old earthquake tracks. Over another saddle the crack goes to Saratoga Creek. Across it and over another saddle it follows Campbell Creek, draining its reservoir. Thence it crosses obliquely the valley of Los Gatos Creek, over the hills of which Bret Harte wrote—

The ridges round Los Gatos Creek
Arched their spines in a feline fashion.

in the earthquake of 1818. Into this creek, from the Feely ranch, some ten acres of land was thrown in a great landslide. At the head of the creek is the long tunnel which cuts under the saddle, from Wright's to Laurel. This tunnel has been the source of endless trouble since it was made, and for the reason that the rock in the mountain through which it passes is made up of minute chips of stone. No wonder, for the earthquake crack follows the mountain ridge, which is here narrow and low. It cuts tunnel and railroad track at right angles, and every earthquake disturbance is sure to make matters worse. Already forty feet of crushed rock has fallen from above what was the roof of the tunnel. On the hill above the tunnel is Morrill's fruit ranch. The earthquake ripped its way through the orchard, shifting the rows of trees six to eight feet and treating roads and fences in the same reckless fashion. The large hospitable Morrill farmhouse stood partly over the track and was split in two and utterly ruined. Farther on at Skylands, on the ridge of the mountains, Fern Gulch was filled with wreckage; redwood trees four and five feet through, two or three hundred years old, were snapped off like whip-lashes. The rift crossed Hinckley's Gulch at right angles. This is a narrow gorge about a hundred feet deep, in which stood the large Loma Prieta sawmill. The gorge was filled by landslips thrown from both sides. The mill was completely buried, with nine mill hands, and a redwood tree over a hundred feet high was set erect and unhurt over the place where the mill stood. The bodies of six men were recovered. One of these, the foreman, was found erect, smothered in mud, but standing with extended arms and limbs in the act of running from the mill. With him, equally erect and in the act of running was the body of a Siberian mastiff. Their position marked the meeting point of the two walls of the cañon. The crack went on across the hills, always in the same direction, southeast by south, till it came to the Chittenden Ranch in the Pájaro Valley. Here it tore off the hillside, destroying the highway at its base; then descended to the Pájaro River, shifting a pier of its railroad bridge about eighteen inches to the northwest. Here it met the Pájaro cross-fault. But here the straight line from Point Arena came to an end. A series of short breaks creeps off to the southeast, ending two miles southwest of San Juan, the last act being the final, almost complete wreck of the beautiful and venerable Mission of San Juan Bautista.

Inverness Road, near Olema, Marin County.

That the oblique crack from Chittenden, famous as an `earthquake ranch' of earlier times, to San Juan, is part of the original rift, is not clear. It may be that this is part of the Pájaro cross-fault. The original Portolá-Tomales fault, if continued in a straight line from Chittenden, would pass along the flanks or the foot of the Gavilan mountains to Priest Valley, fifty miles to the south-southeast. Beyond Priest Valley is a well-marked earthquake crack, which opened in the earthquake of 1868, and in earlier times. This extends through desert land in the same direction, its course being the axis of the Cholame Valley and the uninhabited desert sink known as Carisa Plain. This old rift extends at least one hundred and forty miles beyond Chittenden to Monte Pinos in the north edge of Ventura County. This whole fault from Point Arena to Monte Pinos is clearly a single break, but only 192 miles of a possible 330 were opened in the earthquake of 1906.

Wreck of Loma Prieta Sawmill, Hinckley's Gulch, Santa Cruz County.

But while the surface break seemed to end at Chittenden, it seems probable that the rift in the rocks below extended much farther. At Priest Valley, fifty miles along this line, the shock was violent, while at localities ten miles or more to the east and west of the line, as at Lone Oak Spelled Lonoak by the economical and iconoclastic Postoffice department or the Pinnacles, it was very little felt. In Priest Valley chimneys and shelving were thrown down, buildings badly shaken and the contents of a country store impartially scattered over the floor, the shock being apparently about as severe as in San Francisco.

With the opening of the great rift it is conceivable that faults in the neighborhood should also be affected. There is some evidence (most of which the writer has not examined) of the opening of a parallel fault behind Cape Mendocino. This seems to have passed across the base of the cape, cutting across the smaller headland

Site of Loma Prieta Sawmill, Covered to the Depth of 125 Feet.

called Point Delgada, losing itself in the Sonoma Valley to the southwest of Santa Rosa. There are distinct traces of great disturbance across Burbank's famous orchard at Sebastopol, but it is not clear that in any of these the underlying rock is really broken. Here on a slope lines of fruit trees were shifted, a well was moved bodily three or four feet, and a crack about one fourth mile long extended across a neighboring field, its direction parallel with that of the Tomales rift. Other similar cracks open at intervals on the road toward Point Delgada. At Sobrante, in Contra Costa County, east of San Francisco Bay there are large slumps or cracks in the earth. The extreme violence of the shock in Santa Rosa perhaps indicates its nearness to this second rift, as the Tomales rift caused little damage in other towns equally far away. In some maps of the earthquake rift it is marked as swinging to the eastward in a curve across Point Delgada to the eastward of Cape Mendocino, between that Cape and Humboldt Bay. It seems to the present writer far more probable that the Point Delgada fault is a separate rift, parallel with the main rift, and similar to it, except that it is a little less violent. There is some evidence that a fault line at the foot of San Francisco Bay opened for a short distance to the southward of Milpitas. But the soft soil in that region was filled with slumps and cracks due to the shaking down of loose deposits, and one could not be sure that the actual fault in the rocks was really disturbed. The same remark applies to the breaks at San Bruno about ten miles south of San Francisco in marsh deposits. It is readily conceivable that a great disturbance like the one in the main fault might be accompanied by similar breaks in parallel or associated faults. Or it may be that all these breaks like those in the streets of San Francisco are due wholly to the slump of loose or wet ground.

The studies and experiments made by Professor Branner on the effect of shocks on dry and wet earth leave no doubt that earthquakes are much more severe when the soil is saturated with water. The earthquake of April 18 came at the end of the rainy season, and after a period of great wetness. The soil of the lower grounds along the Bay and in the neighboring valleys rested on gravel saturated with water, and this wet ground was thrown into visible waves by the shocks of the earthquake. On the hills and on rock foundation the shock was intense and sharp, but without these

Old Bogoslof or Castle Island.

destructive surface undulations. In midsummer all the ground would be relatively hard and the injury to buildings would be very much less, considering shocks of like degree of original intensity.

The chief center of disturbance in the earthquake of 1906 would seem to be in the sea. The evidence for this lies in the fact that at the point where the fault enters the land near Point Arena the displacement is greater than anywhere else. As the land fault is traceable for nearly two hundred miles to the southward, it is reasonable to suppose that the sea-bottom is broken for at least an equal distance to the northward. The point of earthquake disturbance off Cape Mendocino has been frequently noticed in the past, and this is in a right line with the rest of the fault. It is possible that the center of trouble is located in the valley between Cape Mendocino and the off-lying submarine mountain.

There is also another possibility, very remote perhaps, but still worth considering, that is, the connection between this rift and the disturbances about the islands of St. John Bogoslof, in Bering Sea. In the southern portion of Bering Sea, about thirty-seven nautical miles northwest from the island of Unalaska, lies a group of small volcanic islets known as Bogoslof in Russian, Joanna Bogoslova, St. John, the Theologian. There are now three of these, all of which have risen from the sea, hot and steaming, within historic times. An especial interest attaches to them just now from the fact that the third and largest of the group appeared at about the time of the great earthquake of April 18, 1906. So far as known its rise must have taken place in March, 1906.

Fire Island, One of the Old Bogoslof Islands.

The possibility of a connection between the disturbances at Bogoslof and those which caused the California earthquake is heightened by the fact that the great earthquake rift, which extends through the Coast Range of California for a distance of 200 miles, follows a direction, which, if produced northward to Bering Sea, would pass near the islands of Bogoslof. Again this earthquake rift was largest, and its effects more violent, where it entered the sea in Mendocino County than at any other point throughout its course.

In opposition to this view may be placed the improbability that an earthquake rift or fault would extend so far as from the center of California to Bering Sea, a distance of more than 2,000 miles, and through such great depths of water as intervene between Point Arena and Bogoslof. It is also stated that the evidence of the seismograph, so far as understood, favors the idea that the great earthquake was confined to California.

Fire Island, One of the Old Bogoslof Islands.

It is evident also that the rise of the third Bogoslof was attended by little if any disturbance in the immediate vicinity. The advent of each of the other two islands was marked by earthquake shocks, the fall of volcanic ashes and displays of fire, observed and felt by the people of Illiuliuk on Unalaska Island. The people of this village in 1906 were unaware of the presence of the new island until the news was brought in by vessels touching at the harbor. Earthquake shocks lasting 30 seconds are reported for May 20 and 23 by the keeper of the light at Scotch Cap on Unimak Island, and a `pretty severe shake' occurred at Dutch Harbor on June 2, but nothing is reported for April or March, when the new island must have risen.

Certainly there could not have been any activity displayed by Makushin or Akutan, both of which volcanoes overlook Unalaska and Dutch Harbor, without being observed by the people of these villages. Perhaps the rise of such an island, in a more or less plastic condition, as it must be, would not necessarily be attended by disturbance in the solid crust of the neighboring islands. On the Pribilof Islands, which had an origin similar to that of the Bogoslofs, no earthquake shock or other disturbance was noted, although these islands were affected at the time of the rise of New Bogoslof in 1883. The Pribilof group lies 120 miles to the north of the Bogoslofs.

On the whole, however, the weight of evidence at present seems to favor the idea that the Bogoslof disturbance of 1906 was local in character and the coincidence in date with the California earthquake involves no actual relation between the two phenomena.

The writer first saw the islands of Bogoslof in July, 1896, while en route for the Pribilof Islands in connection with the fur seal investigations. The U. S. Fish Commission Steamer Albatross attempted to land the commission on Old Bogoslof, but was prevented by the heavy surf, and the thick weather made only a partial view of the islands possible. The vessel afterwards passed the islands on its way to the Commander Islands under more favorable conditions. Dr. Stejneger of the commission obtained some excellent photographs. The writer, still later in the same season, passed both islands while on the British gunboat, Satellite, on the way from the Russian Islands to Unalaska.

At that time Old Bogoslof, known to the sealers as Castle Island, from its appearance, was cold and dead. It showed in the fog a sheer cliff or hill of ashes about 300 or 400 feet in height, seeming much higher in the uncertain light. It was apparently the home of countless sea birds and a small herd of the gray sea lions (Eumetopias stelleri) was hauled out upon one of its slopes.

About half a mile to the northwest lay the islet of New Bogoslof, of about twice the height of the other and considerably greater area. This island was locally known as Fire Island, having but recently ceased to steam and smoke. There was in 1896 no evidence of activity in it, but the water was said to be still warm in the crevices of the rocks. The name Grewingk, in honor of the Russian geographer who compiled an early account of Old Bogoslof, has been given to this island by Mr. Dall.

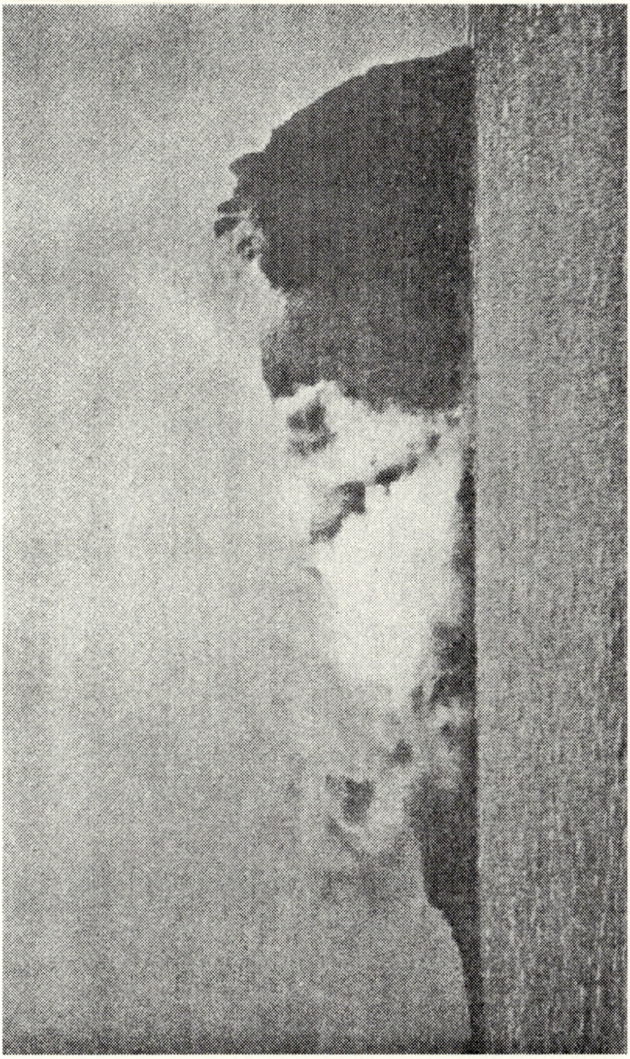

The New Bogoslof Island.

Bogoslof of May, 1906. From New Bogoslof, or Fire Island.

Both islands were surrounded with deep water. In fact the space occupied by the second island had formerly been safely traversed by vessels.

Dredge hauls by the Albatross about the islands resulted in the taking of a number of deep sea forms of fishes, among them three 'grenadiers' (Albatrossia pectoralis, Bogoslovius clarki, and Macrourus cinereus). These were obtained at a depth of 664 fathoms or 3,984 feet.

The Three Bogoslofs, May, 1906.

Conspicuous in the group of islets was an isolated pillar of rock, of considerable height, known as Ship or Sail Rock. It had existed from the earliest times, having been reported as early as 1768. It was seen by Captain Cook in 1778, who mistook it for a ship under sail, hence its name. This was eighteen years before the rise of Old Bogoslof. Ship Rock crumbled and fell in ruins about 1888.

About April, 1906, midway between Old and New Bogoslof, a third island, larger than either of the others appeared. Captain Dirks of Dutch Harbor estimates its size as five times that of New Bogoslof, although the photographs do not seem to bear this out. This new island was first seen by the U. S. Fish Commission Steamer Albatross, Captain L. M. Garrett, on May 28, 1906, while on her way to the investigation, under direction of Professor Charles H. Gilbert, of the fisheries of Japan. Soon after this date the U. S. Revenue Cutter Perry visited the islands. Photographs taken by officers of the Perry and supplied by Mr. H. H. Taylor, of the North American Commercial Company, are here reproduced, together with photographs of Castle Island and Fire Island, taken by Mr. N. B. Miller of the Albatross, in 1892.

The early history of these very interesting islands is given by Professor George Davidson in the *Bulletin of the American Geological Society, Vol. XXII.*, p. 267, and a detailed and exhaustive account of them by Dr. C. Hart Merriam, profusely illustrated, appears in the Report of the Harriman Expedition of 1899, Vol. II., p. 291-336

Of the advent of the first island, in 1796, the following account is given in Kotzebue's narrative of discovery in 1817. The story is that of a Russian trader, Kriukof, who found himself with some native hunters forced to seek refuge from storm on the north end of Umnak Island, the island of the Aleutian chain, nearest the Bogoslofs. It was in May and when the storm cleared on the 8th, Kotzebue tells us:

They saw to the N., several miles from land, a column of smoke ascending from the sea; toward evening they observed under the smoke something black, which arose but a little above the surface of the water. During the night fire ascended into the air near the spot, and sometimes so violent, and to such height, that on their island, which was ten miles distant, everything could distinctly be seen by its light. An earthquake shook their island, and a frightful noise echoed from the mountains in the S. The poor hunters were in deadly anxiety, the rising island threw stones towards them, and they every moment expected to perish. At the rising of the sun the quaking ceased, the fire visibly decreased, and they now plainly saw an island of the form of a pointed black cap. When Kriukof visited the island of Oomnak, a month afterward, he found the new island, which during that time had continued to emit fire, considerably higher. After that time it threw out less fire, but more smoke; it had increased in height and circumference and often changed its form. For four years no more smoke was seen, and in the eighth year the hunters resolved to visit it, as they observed that many sea lions resorted to it. The water round the island was found warm, and the island itself so hot in many places that they could not tread on it.

The eruption of 1883, which resulted in the rise of New Bogoslof, seems to have had no eye-witnesses and the exact date of its appearance is unknown. Captain Anderson, of the schooner Matthew Turner, saw the new island in September, 1883, and reported that great volumes of steam and smoke, accompanied by showers of ashes, were thrown out from the summit and through fissures in the sides and base, the bright reflections from the heated exterior being visible at night. At the time of this eruption a severe earthquake was felt in the sea off Cape Mendocino, apparently in the line of the Portolá-Tomales rift of April, 1906.

The islands were visited in 1884 by the officers of the U.S. Revenue Cutter Corwin, and Lieutenant J.C. Cantwell and Surgeon H.W. Yemans made the ascent of New Bogoslof. Lieutenant Cantwell thus describes his experience in the 'Cruise of the Corwin':

The sides of New Bogoslof rise with a gentle slope to the crater. The ascent at first appears easy, but a thin layer of ashes, formed into a crust by the action of rain and moisture, is not strong enough to sustain a man's weight. At every step my feet crushed through the outer covering and I sank at first ankle-deep and later on knee-deep into a soft, almost impalpable dust which arose in clouds and nearly suffocated me. As the summit was reached the heat of the ashes became unbearable and I was forced to continue the ascent by picking my way over rocks whose surfaces, being exposed to the air, were somewhat cooled and afforded a more secure foothold.

On all sides of the cone there are openings through which steam escaped with more or less energy. I observed from some vents the steam was emitted at regular intervals, while from others it issued with no intermission. Around each vent there was a thick deposit of sulphur which gave off suffocating vapors.

The islands were visited by Drs. C. Hart Merriam and Thomas C. Mendenhall of the Bering Sea Fur Seal Commission in 1891. Dr. Merriam writes thus of New Bogoslof as seen at that time:

The new volcano was enveloped in steam, which issued from thousands of small cracks and crannies and poured in vast clouds from a few great fissures and crater-like openings, the principal of which was near the northwest corner, only a few feet above high-water mark. From this opening, the shape of which we could not see, it rushed out with a loud roaring noise. So great was the quantity of steam that it completely concealed the upper part of the island except when wafted to and fro by violent gusts of wind. . . . The steam was usually impregnated with fumes of sulphur, and deposits of sulphur, some in very fine needles, were observed along the margins of the cracks.

Of the third Bogoslof, Dr. Charles H. Gilbert, of Stanford University, who was in charge of the work of the Albatross when the `brand new mountain' was first seen on May 28, 1906, writes thus in a personal letter regarding it:

When I saw it (Bogoslof) in 1890 there were really two small islands about 1½ miles apart, one of them steaming and the other already cooled off. This has been the condition for a number of years, so the hot one had received the name of Fire Island, the cold one, Castle Island. When they came in sight yesterday, we were astonished to find that Fire Island was no longer smoking and that a very large third island had arisen half way between the other two. It was made of jagged, rugged lava and was giving off clouds of steam and smoke from any number of little craters scattered all over it. Around these craters, the rocks were all crusted with yellow sulphur.

In a later letter, written from Yokohama, Dr. Gilbert said:

I wrote you a full account of Bogoslof, but the letter seems to have miscarried. Our discovery seems to have been corroborated later by some revenue cutter, but if the newspaper report agrees with their findings, very extensive changes took place in the interval between the two visits. When seen by us, the new cone, occupying much of the space between the two older ones, was somewhat higher than either, but was certainly far from 900 feet high—300 feet would be an extreme figure. There was no evidence of a central crater. The steam and fumes were given off most abundantly from cracks and fumaroles on the slopes. About these were heavy incrustations of sulphur. We saw no indications of boiling water, nor did we believe that landing would be impossible.

In an account of the physical history of the Bogoslofs, written in 1899 for the report of the Harriman Expedition, Dr. Grove K. Gilbert, of the U.S. Geological Survey, noting the rapid disintegration of the islands, said:

One might predict that in the next century the name Bogoslof would attach only to a reef or shoal, were it not for the possibility of new eruptions. The pulse of the volcano is so slow that we have noted only two beats in more than a century, but such sluggishness must not be taken as a symptom of death, or even decline, for volcanic organisms are characteristically spasmodic in their activity. Long before the sea has established its perfect sway the arteries of the mountain may again be opened and a new and larger island put forth to contest its supremacy.

Nearly a century elapsed between the arrival of the first and second Bogoslof, only twenty-three years between the second and third.

The floor of the depths of Bering Sea in this region seems to be still unsettled, and astonishing changes may be looked for at any time. If it should prove true that the geological faults of California extend out from this center, a new interest would be attached to the outbreaks of Bogoslof.

In the fall of 1906, after the close of navigation, in Bering Sea, according to Mr. H. H. Taylor, of the North American Commercial Company, two violent shocks were felt at Unalaska. The people of that island are waiting with interest to see what new changes have taken place in the unsettled Bogoslof.

The earthquake of 1906 is receiving the most thorough study possible, and in such a way as to give promise of important practical results. The State of California has formed an Earthquake Commission consisting of geological experts, and these have received in their work important financial assistance from the Carnegie Institution. The details of the earthquake rift and the effects of the shocks on buildings have been carefully recorded and photographed. The final report of this commission should leave no important question in doubt. Many previous earthquakes have been recorded in California, but their most essential feature, the location and extent of the causing fissure has rarely been indicated. In the records we read again and again that `fissures opened in the ground,' but whether these were rifts in the crust or mere slumps of soft ground as a rule has escaped attention. The great earthquake of 1868 opened rifts at intervals from Tomales Bay to Carisa Plain, and also a fissure on the east side of San Francisco Bay, where a straight crack about ten miles long extended from Haywards toward the south. One side of this rift showed a lateral displacement of about four feet. To this short rift, rather than to the Portolá-Tomales fissure, the shock in San Francisco in 1868 may have been due. The shock in that year was more violent in Oakland than in San Francisco and most violent about San Leandro and Haywards, to the south of Oakland. It is conceivable that the shock of 1865, having its center in the Portolá fault, not far from San Francisco, gave that city a degree of immunity in 1868. Other destructive earthquakes, as recorded by Holden ('Catalogue of Earthquakes on the Pacific Coast, 1769 to 1897') are as follows:

1800. This earthquake was severe about San Juan Bautista, but whether in the Portolá fault or the Pájaro fault is not clear.

1812. This earthquake wrecked the mission of San Juan Capistrano in southern California, and was felt along the line of the southern missions. It had its center possibly in the Santa Catalina Channel.

1818. This earthquake injured the mission of Santa Clara; hence it may have been along the Portolá fault. All the houses in the Santa Clara Valley were shaken down at about this time.

1836. This was said to be similar to the shock of 1868, its center along the Portolá line; great fissures were made in the earth.

1839. This was severe from Redwood to San Francisco, a great fissure opened to Mission San José. It was probably also in the Pájaro fault.

1857. Sacramento to Fort Tejón, San Bernardino and Fort Yuma. At Fort Tejón a fissure 20 feet wide and 40 miles long: the sides came together with such violence as to make a ridge ten feet wide and several feet high. Fissures at San Bernardino.

1865. This was a smart shock from San Francisco to San José, apparently along the line of the Portolá fault. The severity of this earthquake, as suggested above, may have mitigated the local severity of the earthquake of 1868, which was in the same rift, but not so severe in this part of it.

1867. This was violent disturbance about Klamath Lake. A great crack said to have opened in Siskiyou County, but the locality is not recorded.

1868. A very severe earthquake, there being a rift on the east side of the bay, as also at Olema, in the Santa Cruz Mountains and for over a hundred miles from Cholame through the Carisa Plain.

1872. Owens River, Inyo County. Fissure at Big Pine 50 to 200 feet wide, 20 feet deep, extending 50 miles or more. Numerous shocks, very violent, these preceded by weaker shocks for a year or more. It is said that in the rift of this earthquake, still open, may be seen the mummified cattle which were engulfed in it in 1872. This statement is given on the authority of Mrs. Mary Austin.

1890. Mono Lake, similar disturbances.

1892. Vacaville, Winters, etc., extensive local disturbances, on the west side of the valley of Solano and Yolo. This rift extends through the volcanic basin of Clear Lake to the northward, parallel with the Tomales fault.

1897. San Jacinto Valley, with a notable fissure, the details not at hand.

To these might be added the vigorous single jolt of 1893 in the San Fernando Mountains, which did little harm because occurring in an uninhabited region. The writer was at Saugus at the time, and noted the fall of trees and the flinging of rocks down the mountainside. There seems to have been but a single wave, which would have done great injury in a populous district. It is said to have arisen from a short fissure in Pico Cañon.

Since the earthquake of 1906 many small earthquake waves have followed, evidently harmless details in the process of adjustment. Looking over Holden's record, we see that many small disturbances have taken place along the line of the great fault in question, besides the great earthquakes of 1868 and 1906 and the lesser ones of 1800, 1818, 1836, 1839, 1865 and 1868.

In 1808 there were twenty-one shocks at the Presidio of San Francisco. In 1812 the shocks caused a tidal wave in the bay extending up to the plaza. In 1813 or 1815 'all the buildings' in Santa Clara Valley were shaken down. There were not many and all these were of adobe or sun-dried brick. In 1851, a sharp shock in San Francisco. In 1852, a shock at San Francisco, with a fissure, through which Lake Merced drained into the sea.

1853. Heavy shocks near Humboldt Bay.

1856. Severe shocks at San Francisco, the water in the bay sank two feet.

1863, 1864. A sharp shock at San Juan Bautista.

1890. Sharp shock along Portolá fault. The Pájaro bridge had a pier shifted 18 inches, as in 1906. The same crack opened at Chittenden, and the main arch in the Mission Church at San Juan Bautista was

> injured. A rift opened in the soil from Chittenden to San Juan as in 1906.

In a general way this seems to be proved without question: The great earthquakes in non-volcanic regions occur always in the same rifts. They also occur with a certain sort of periodicity. The Portolá rift with its destructive earthquakes of 1812, 1836, 1868 and 1906, seems to have a period of thirty to forty years. In the intervening time the region of this rift may be reasonably regarded as immune.

The same sort of periodicity, with an interval of about 30 years, has been noticed in the seven earthquakes of Chile.

These great semi-periodical earth-changes are known in Spanish speaking countries as terramoto. It is the minor waves which the Spaniards call temblor.

We may again emphasize the fact that the earthquakes of California are purely mechanical in their origin. The coast, for some reason connected with the secular changes in continents and seas, is in California and along the whole run of the Pacific slowly rising. Perhaps some part of the Pacific Ocean is slowly sinking, as part of the same movement. In any event, the rise of this heavily loaded and much broken area is accompanied by a heavy strain. This the break in the rocks apparently relieves, and when the strain reaches a certain point it will break again, and this at more or less regular intervals.

The earthquake at Valparaiso seems to have been caused by the same kind of a strain operating in the same way. As a final result of it the whole coast of Chile has been elevated.

While many slumps and breaks in the land have been noted in Chile, there is no continuous rock-rift or fault like that shown by the California earthquakes. Undoubtedly the rift was in the sea, and probably parallel with the line of the coast.

The tremendous earthquakes of a few years ago in Alaska, along the foot of Mount St. Elias and Mount Fairweather, earthquakes which wrecked the Muir Glacier and altered that of Malaspina must have been of the same nature and origin. Similar strains on the same strained ocean-rim probably caused the Bogoslof disturbances, and Jamaica is not so far from the Pacific that it may not be included in the same circle of disturbance. The Asiatic side of the Pacific is still more unsteady as a foothold for man. Japan, Formosa, and the Philippines are notoriously subject to earthquake rifts as well as to volcanic outbreaks. The islands of the straits of Sunda have a world-wide reputation in these regards, and if press statements are accurate, there have been of late great disturbances in the Solomon Islands. As most of the islands of Polynesia and Micronesia are volcanoes, or of volcanic origin, disturbances in these regions need not cause surprise.

In the California earthquakes there are no explosions, no bursting of 'caves of gloom,' no escape of sulphurous or other gases, and none of the phenomena real or alleged which excite the imagination of the superstitions of the ignorant or credulous. They are not caused by electricity nor accompanied by electrical disturbances. Electricity is a mighty force only when held in great tension by some form of insulation. Land and sea, unlike storm clouds, offer no means of electrical insulation, and electrical tension cannot be accumulated along the sea bottoms where so many earthquake disturbances occur.

The electrical theory of earthquakes is thus intelligibly given by Mr. Charles Hallock, the account being a condensation of his views as stated in an article called 'Polarity of the Seismic Impulse':

Accepting the theory of Sir Oliver Lodge and other advanced scientists of eminence, that the earth is a magnet, and its crust simply the armature of an immense dynamo whose source is the sun, Mr. Hallock attributes the unusual prevalence of earthquakes, and the manifest sympathy between so many of them, to the earth being surcharged with electricity. This redundant voltage sets in motion the loose heterogeneous masses of which the terrene envelope is so largely composed, and these in turn generate electrical energy. This is in line with Clerk Maxwell's showing, the rock and earth movements being facilitated by excessive rains which saturate and lubricate the dislocated masses. Jarring is often maintained for weeks and months at intervals, after the initial shocks, in obedience to the law of adjustment which causes disintegrated particles to settle and become compact.

Mr. Hallock goes on to aver that all our troubles and calamities are not due to natural causes alone, but that men's dabbling with electricity, to the extent of gridironing nearly the entire globe with wire conductors, overhead and underground, in the atmosphere and through the oceans, and in practically every house in town and country, and keeping them constantly charged by powerful dynamoes, is the stimulating cause of current perturbations, landslips and volcanic eruptions, which are far in excess of any known period since the Tertiary. All of which he takes occasion to remind us is predicted in the book of Isaiah, which declares that 'men shall be plagued by their own inventions.'

It is clear that this view does not account for the conditions in California. The great earthquake rift follows the line of a fault which shows clearly traces of many earthquakes before any men lived in California. There is no evidence that the recent earthquake was any more severe than those of 1836 or 1813. It is much less severe in San Francisco than anywhere along the rift in Marin or Mendocino Counties. It was even less severe in San Francisco than along the possible secondary rift in Sonoma County. It is perfectly clear that the operations of man at San Francisco had nothing to do with the rift of rock in Marin and Mendocino Counties, nor with its extension in 1868 along the line of Cholame and the Carisa Plain.

No one can check an earthquake or modify its action. There is no lightning-rod insurance against it. Fortunately also no one can set it off or stimulate it to any greater violence than nature has intended.

There is no reason to suppose that any planetary conditions produce earthquakes. The conjunction of the planets, even all of them, would produce less variation in strain than the conjunction of the sun and moon which occurs every month. That is adequate to produce variation in tides, but scarcely enough to be detected by the finest balances. A force too small to be weighed is scarcely likely to shift mountains, though it is conceivable that it might set them off, under conditions like that of the proverbial last straw on the overstrained camel. There is no evidence of connection between earthquakes and sunspots. They may appear coincident. The sun is seldom without spots and there are few years without a destructive earthquake on some part of the earth's surface. Little tremors of one sort or another occur somewhere or other every hour, and the crust of the earth is never quite at rest.

There is no evidence of connection of earthquakes with any kind of climatic condition. The notion of 'earthquake weather' is an absurd superstition. Earthquakes have no preference for any month, for any time of the day. Wet ground slumps more than dry ground, but the wetness is not a cause of the earthquakes. They cannot be predicted except in the most general way, as I might say, the next earthquake at San Francisco is due in forty years, namely in 1946. Soothsayers and clairvoyants being vulgar, ignorant and predatory persons of criminal instincts know not more but less about earthquakes than the average decent citizen.

Moreover, it is no longer believed that the wickedness of man produces earthquakes. This has its own reward, but the sin and the penalty are like in kind. The mountains break as an overweighted leaf of a table would break when the strain is growing greater and it can be borne no longer.

The earthquake of 1868 was far less violent than that of 1906, along the San Francisco peninsula, although it extended farther to the south than the other. It may be remembered that the population of the region is now much greater than in 1868, and in like manner, the possibilities of mischief on the part of earthquakes has been correspondingly increased. The danger from earthquake itself is relatively a small matter, but it should be considered in the building arrangements of regions where such disturbances are likely to recur. It is as easy to make buildings virtually earthquake-proof as water-proof, unless standing directly over the fault itself. Earthquake-proof structures can be made of steel or of properly mixed concrete, properly reinforced by steel, or within certain limits of brick or stone firmly cemented and with roofs properly made and properly tied together. Brick and mortar are not adequate for the existing conditions of California. A firm foundation is moreover of primary importance. In loose soil, and especially in sand, earthquake waves are much higher, longer, slower and more destructive than in rock. In this connection we may close with the pertinent words of the engineer, William H. Hall, of San Francisco: "The earthquake has put a definition on the word sham, which seems positively cruel. It has established a value on the solid foundation and genuine superstructure which is indeed ennobling."

It would redound to the moral and spiritual elevation of any community to be assured of a smart shock or temblor at intervals and of a real earthquake or terramoto once in each generation.

Geology and the Earthquake

By John Casper Branner

Vice-President and Professor of Geology at Stanford University

What is it? What causes it? Where does it come from? Will it happen again? and, if so, when and where and how much? These are the questions the geologist is expected to answer regarding earthquakes in general, and in particular regarding the California earthquake of April 18. And, as usually happens in such cases, the geologist can half answer some of these questions, and others he can not answer at all.

To begin with the last item—the "how much?" Was the late earthquake really a severe one, or was not its severity and importance greatly exaggerated as compared with great earthquakes, and was not this exaggeration carried still further by the burning of San Francisco, which immediately followed?

The scales in use among geologists for classifying earthquakes divide them into ten classes, according to their violence or intensity. By the Rossi-Forel or the Mercalli scales, the California earthquake stands somewhere between eight and ten at points of greatest disturbance; from which we infer that we may trust our senses to the extent of believing that it was no small affair.

The picturesque and sensational features of earthquakes are abundant and entertaining, but to the geologist these features have only a passing and accidental interest. For example, if a chimney top, broken off by an earthquake, should fall on a man in such a fashion as to go right over his head and leave him standing unhurt in the flue, it would be a striking, and to the man a very important,

Slump in Soft Ground, Milpitas.

fact; but, from a geological point of view, its only importance would lie in the fact that the shock was severe enough to throw down the chimney. One hears, in the vicinity of Palo Alto, of a herd of cattle having been swallowed up in the Santa Cruz Mountains and how they had to be dug out. This seems like a genuine earthquake tragedy; but it turns out that at the time of the earthquake there was a landslide affecting some ten acres of land on which the herd was grazing. On the higher side, the slide left banks up which the cattle could not climb, so that a road had to be dug to get them out. This again turns out to be a matter of but little importance from the geologic point of view. Mention is made of such cases simply to call attention away from the strange and bizarre, and to direct it more effectively to what are regarded as matters of fundamental importance in connection with earthquakes. The phenomena that bear directly upon the causes and throw some light upon the past and future seismic history of the region are evidently the ones of the deepest importance, and it is to some of these that attention is directed.

Rift Crossing Road near Skylands, Santa Cruz County, Showing Relative Sinking of West or Uphill Side of Rift.

In the Portolá Valley, five miles west of Stanford University, there runs through a pasture field what looks like a plowed furrow. It is such a furrow as might be made by a big turning-plow, except that the sod is not turned clear over, the clods and grass roots are rough and irregular, and the furrow not straight or gently curved, but ragged and lumpy and sometimes forked. Where this furrow crosses a fence-line, the fence has an offset in it amounting to a little more than eight feet. A few hundred yards away, it crosses another fence-row at a low angle and here there is eight feet more then is now needed; where it crosses a third fence at a right angle there is another offset in the fence of fully eight feet; at another place it crosses a line of water-mains, and the pipes are displaced more than six feet.

To the passing observer these facts may appear trivial enough, but to the geologist they are full of interest and importance, for they lie close to the source of the earthquake; they are produced by a lateral displacement along a line of fracture in the crust of the earth. Where this fracture passes through the Santa Cruz range of mountains, it was worked out from the geology some ten or fifteen years ago and was gradually traced in detail for a distance of forty-five miles. Starting on the coast at Mussel Rock, seven-and-a-half miles south of the Cliff House, it takes a course of south about 40 degrees east, following certain topographic features that are plain enough on the ground. It runs through San Andreas and Crystal Springs Lakes, Portolá Valley, passes just west of the peak of Black Mountain, follows along Stevens Creek Cañon, and, passing to the west of Loma Prieta, continues in the direction of Sargents Station on the Southern Pacific Railway.

Rift Across Road near Azul Springs, Santa Clara County.

Immediately after the recent earthquake this old line of fracture was visited at a number of places and everywhere it showed evidence of having been newly broken and displaced. The displacement was mostly a lateral one, amounting to a maximum of eight-and-a-half feet, but there was also some vertical movement which probably does not exceed three feet in the region thus far examined. The country southwest of the fault sank and moved at the same time toward the northwest, or else the region on the opposite side rose two or three feet and moved about eight feet toward the southeast. When we reflect that these mountain masses were moved such a distance in a few seconds and stopped suddenly, there is no cause for wonder at the jar produced in the adjacent region.

The materials visible in the line of this fault are worthy of note. The fault is an old one, along which many and great movements have taken place; the rocks have therefore not only been broken across, but they have been crushed, recrushed and ground up until it is now difficult or impossible to find large blocks close to the fault-line. Furthermore, the word "line" is here somewhat misleading, for it is really a belt or zone, from twenty to fifty or one hundred feet across, rather than a clean-cut line or plane. At the surface, decomposition has further attacked the materials and the soil is commonly deep and yielding and this soil has in all probability taken up a good deal of the actual displacement by lagging, stretching and crushing; this seems to account for the fact that the displacement is not everywhere of the same amount.

House over Rift near Saratoga Springs, California.

The phenomena to be seen at the surface along this line of displacement are such as might be expected. Wherever fences cross the fracture at right angles they are torn in two and the broken ends now stand from one to eight feet apart; roads that were formerly straight are now bent; barbed-wire fences are pulled in two or they are variously shortened; water-pipe lines have the pipes either broken and pulled apart, or where the pipe line crosses the fracture at a low angle the pipes are telescoped into each other from four to six feet. A dam across Crystal Springs Lake crossed this old fault-line at right angles; it was expected that the dam would be torn in two or badly fractured, but it was so well built that the fault was compelled to pass round the dam and through the rocks at its end. The same thing happened at the dam across the east end of San Andreas Lake. Where trees stood directly upon the break they have been uprooted, and in some cases they have been split in two.

Such are a few of the evidences of a displacement of the rocks over a distance of some seventy-five miles through the Santa Cruz Mountains south of San Francisco. North of that city the topography suggests that this same, or a closely related, fault passes through Tomales Bay, down the Gualala and Garcia Rivers, entering the ocean at the town of Manchester near Point Arena light-house. It is expected that a further examination will disclose similar evidences of displacement along this line north of San Francisco.

The fault-line mentioned, however, is far from being the only one in the Coast Ranges. The long parallel valleys of the State are due, in part at least, to faulting that took place a long while ago. One great fault, that seems to have been involved in the California earthquake, follows the entire length of the Santa Clara Valley, from about the headwaters of the San Benito River south of Hollister, past San José, through the Bay of San Francisco, up the valley past Santa Rosa, Ukiah, Willits and down the Eel River, or parallel with it, to Eureka in Humboldt County. Since the earthquake, this fault-line has not been seen by the writer, but many cracks have opened along

Redwood Snapped Off by Earthquake near Fort Ross, Cal.

its axis near the south end of the Bay of San Francisco between Milpitas and Alviso. At this place not only were cracks opened from one to four feet wide and five or six feet deep in the soil, but for a couple of days water ran out through some of the cracks bringing up sand and forming small cones about them. Some wells began to overflow that hitherto had never done so, the flow in other wells increased very decidedly, while in still others the water sank somewhat.

Evidently the earthquake and the faults are related; but did the faults make the earthquake, or did the earthquake make the faults? It is a fair question. Look at the matter for a moment from a purely theoretical point of view. Conceive of a mass of rock as big as a big house under pressure enough to break it—would not the breaking produce a jarring of the surrounding mass? Or imagine such a rock already broken across and the two faces forced past each other for a distance of eight feet—would not this movement jar the

The Fault Passes Under a Live Oak and Uproots It, Woodside, Cal.

surrounding mass? And if the break were three hundred miles long would not the jar extend into the adjacent rocks and soil in the same fashion and for many miles? This theory seems to explain the earthquake.

There are fractures, however, that evidently must be attributed to the earthquake; such are those connected with landslides, the disturbance of steep and unstable slopes, the settling of loose masses of gravels and the like in wet ground. But these are all matters of small importance and quite incapable of producing earthquakes, except of a very local kind.

It is plain enough that faults are caused by unequal pressure developed in the rocks. This pressure may bend the rocks, or it may break them and thrust them past each other; and when they break, the fractures may pass down for thousands of feet, or even for miles, into the rocks beneath the surface. What causes this strain or

Rift on Shafter's Ranch, Olema, Cal.

inequality of pressure is not so evident. Three theories have been suggested: first, the cooling and consequent contraction of hot rocks; second, the heating and consequent expansion of cool rocks; third, the shifting of loads upon the earth's crust by the washing of land-masses into the sea.

But whatever theory one adopts regarding the remote causes of earthquakes, the conclusion is inevitable that they are produced by natural causes, one of which is the relief of strains within the earth's crust along lines of fracture. The knowledge that they are due to natural causes ought to contribute to a philosophical view of them and rid them to some extent of the terror they inspire in the minds of those who attribute them to the wrath of God or to other supernatural causes.

As for the earthquake happening again, the only guide the geologist has is the record found in the rocks. This record shows plainly enough that there always have been earthquakes. As for anything more specific in regard to time and place and violence of future earthquakes, the geologist must leave prophecy to the prophets.

The Destructive Extent of the California Earthquake. Its Effect Upon Structures and Structural Materials Within the Earthquake Belt

By Charles Derleth, Jr.

Associate Professor of Structural Engineering, University of California

On the eighteenth of April, 1906, at 5:13 in the morning, one of the most severe earthquakes ever recorded since the beginning of civilized habitation visited the State of California. The important destruction to engineering works occurred in a belt about fifty miles wide and nearly three hundred miles in length, extending along the Pacific Coast, with the Bay of San Francisco at its center.

Introduction

Immediately after the earthquake the Governor of California, the Hon. George C. Pardee, established the California Earthquake Investigation Commission, which has diligently studied the scientific phase of the subject. The commission's preliminary report clearly outlines the general geological features of the earthquake belt and the important phenomena observed in the region of most violent shock. The commission has carefully studied time records to determine coseismal curves, and collected data for the purpose of establishing lines of equal severity or intensity of shock, that is, isoseismal curves. As one of the conclusions to their work the commission will undoubtedly discuss the relation of earthquake effects to man and to human works, that is, engineering construction; but as that body consists mainly of pure scientists, namely, geologists, physicists, and astronomers, their work in the interpretation of the destruction to structures can not be complete without the cooperation of students of engineering.

Members of the San Francisco Association of the American Society of Civil Engineers have considered the engineering side of this large problem, and various sub-committees have reported detailed studies for different and distinctive types of construction, such as buildings, streets, harbor works, water systems, sewers, railroads, power stations, and foundations. These reports have been forwarded to the parent society in New York and have been printed in the March (1907) Proceedings of the American Society of Civil Engineers.

A local association given the name of the "Structural Association of San Francisco," consisting of engineers, architects, contractors and manufacturers of structural materials, with a purpose similar to that of the American Society of Civil Engineers, is also in the field. Its members, too, have appointed sub-committees for a division of labor in collecting and studying data in this vast problem.

The Structural Association is confining its attention mainly to San Francisco and the question of earthquake-proof and fire-proof buildings.

At this writing a report is in press, written for the United States Geological Survey by Messrs. John S. Sewell, Frank Soulé and Richard L. Humphrey, in which these gentlemen, all engineers, treat of the effect of earthquakes and fire upon structural materials.

Furthermore, many articles by engineers have appeared during the past year in the engineering journals, treating of the earthquake and fire problems. It is plain, therefore, that the subject is receiving considerable attention by those interested in building and in structural materials.

Engineers are concerned not only with the temblor's destruction, but also with the fire problem. Immediately after the earthquake, great conflagrations broke out in San Francisco and Santa Rosa. The earthquake meted out great destruction, and the large losses will be felt for some time to come; but there is always at least a little good accompanying evil, and all intelligent and honest builders have recognized that the calamity offers a great opportunity to compare the efficiencies of different types of design and to observe the relative behavior of different kinds of materials in their resisting qualities to withstand earthquake shock and retard the progress of fire. It seems to me that the engineering problem is at least as large as that of the geologist; at any rate it is more important in its practical bearings, because it combines the study of structural stability with the theory of fire-proofing and must pay considerable attention to the relation of destruction of structures to geological formations.

From the purely scientific standpoint this earthquake presents perhaps the most favorable problem which it has yet been the privilege of seismologists to study, because the extent of the earthquake is so large, the area of destruction embraces such varied topography, and because the geological formations of the Pacific Coast are so striking and so unique. From the very first, the center or line of disturbance has not been in doubt, for a crack is visible on the earth's surface for at least 200 miles, and runs in an almost unbroken straight line along an old geological scarp. This scarp, or plane of crustal weakness, is plainly visible to the educated geological eye, and has been known to geologists for more than a generation. Again, the magnitude of the shock was so considerable that its vibrations were felt at many places quite remote from San Francisco. The tremors were distinctly felt in the southern part of California, in Oregon, and at several places in Nevada; while precise instruments have recorded small crustal movements at Washington, D.C., in Germany, and at Tokyo. We have here a fruitful opportunity for an advance in the world's knowledge of geophysics, and scientists generally will look with anticipation to the final report of the California State Earthquake Commission.

For the engineer, from the purely applied science point of view, there is an equally wide opportunity. All kinds of construction and all kinds of material have been subjected to both stress and fire. Structures, good and bad, of able and deficient designs, of honest and criminal workmanship, all have been tested by various degrees of vibration, from the most severe shocks in the region of the fault line to shocks of much less severity for places resting upon firm foundations.

Earthquakes and Crustal Movements

The crust of the earth is constantly adjusting itself to conditions of stress and strain. The surface of the globe is gradually and slowly changing its form to suit these adjustments. The span of a human life is quite negligible in comparison to the geologic ages required to bring about marked deformations in the surface of the globe, and consequently many of us are not aware of the slow crustal movements which to the eye of the experienced geological observer are everywhere in evidence upon the world's surface.

Some parts of the earth's crust are slowly sinking; a portion of the east coast of the State of New Jersey is said to be dropping. In other places the land is rising; it is claimed that the coast ranges in California are young mountains pushing their way through the coastal plain. Many such statements for different parts of the surface of the globe may be cited. These are effects of the so-called mountain-making or tectonic forces which act through long periods of time and over wide areas.

Some parts of the earth's surface are more settled or stable than others, and we do not expect severe crustal movements in such regions. New York City, resting on strong elastic rocks, is probably an example. The Adirondacks and vicinity, whose foundation is of the earliest geologic age, is never associated with earthquakes. In other places, crustal movements, that is earthquake phenomena, are from time to time to be expected. Japan and the Pacific Coast of America are such countries. As the earth's crust gradually changes, lines of weakness no doubt will be shifted from one part to another of the globe's surface, and what are now termed "earthquake countries" may no longer be so in the next geologic age.

I must distinguish between the volcanic and tectonic earthquakes. In the present discussion it is not necessary to consider the volcanic type. Volcanic earthquakes are generally local and their effects of smaller extent. They are comparatively rare. The tectonic or mountain-making earthquakes are more frequent in occurrence; they may affect large surfaces on the globe, as in the present instance; they may be severe, or so slight that only the most delicate seismic instruments will detect them. For a year previous to our great April earthquake, shocks were recorded by delicate instruments with great frequency in the neighborhood of San Francisco. While tectonic earthquakes are apt to occur more often in regions of proved crustal weakness and instability, they are nevertheless liable to occur anywhere and at any time.

The crust of California, dynamically speaking, is alive and active. Here the earth's surface is in growth and we are witnessing one instant perhaps in its development. Geologically speaking, the earth's crust in California is somewhat unsettled or unstable, and I see no object to be gained by not admitting this fact. Earthquakes are natural phenomena and should not be feared. We can not contend with nature's forces, but we certainly can try to adjust ourselves and our works most favorably to their requirements. Earthquakes offer to the geologist most interesting dynamical problems, and their effects upon human construction, no matter how unfortunate, lend to the engineer and the artisan most valuable experience and counsel.

Fault Lines in California

More than a generation ago geologists mapped out long continuous fault lines upon the face of California. These lines are the results of former slippings or accumulations of slips that have occurred in the past, often the remote geological past, and thorough study reveals

FIG. 1—*Relief Map of San Francisco Peninsula, by Professor Andrew C. Lawson; Consult Fifteenth Annual Report, U. S. Geological Survey, Page 405.*

thei

r relation to California topography. Most of these important fault lines, if not all, run in a general north-north-westerly direction, essentially parallel to the mountain range lines, a parallelism which is quite natural and to be expected. A study of the report of Mr. G.K. Gilbert, Monograph 1, U.S. Geological Survey, describing the phenomena at Lake Bonneville, and the earthquake destruction in Inyo County, California, in 1872, also the work by Professor Andrew C. Lawson entitled "Sketch of the Geology of the San Francisco Peninsula," Fifteenth Annual Report, U.S. Geological Survey, page 405, will fully acquaint the reader with the geology of California and help him the better to appreciate what happened on the eighteenth of April, 1906.

The great earthquake of 1872 was accompanied by and was coincident with heavy slipping along parts of a pronounced fault line which traverses the western flank of the Sierras from Owens Lake in the south toward Lake Tahoe in the north. A relief map of the San Francisco peninsula given by Professor Lawson in the work above referred to, see Fig. 1. , shows clearly a part of another great fault line which follows the coast and runs in the same general direction as the one in the Sierras. This great coast fault is clearly shown by the map, Fig. 1, to run through Lakes San Andreas and Crystal Springs of the Spring Valley Water Company. There are a number of lesser faults to be studied in the coast range region. These faults are lines of weakness in the crust along which movements and slippings have occurred in the past; and renewed ruptures or movements in the rocks far below the surface, at the same time that they produce earth vibrations, may also cause surface cracks and other evidences along these geological scarps,—effects which were so pronounced in 1872 in Inyo County, and this time in the coast ranges near San Francisco.

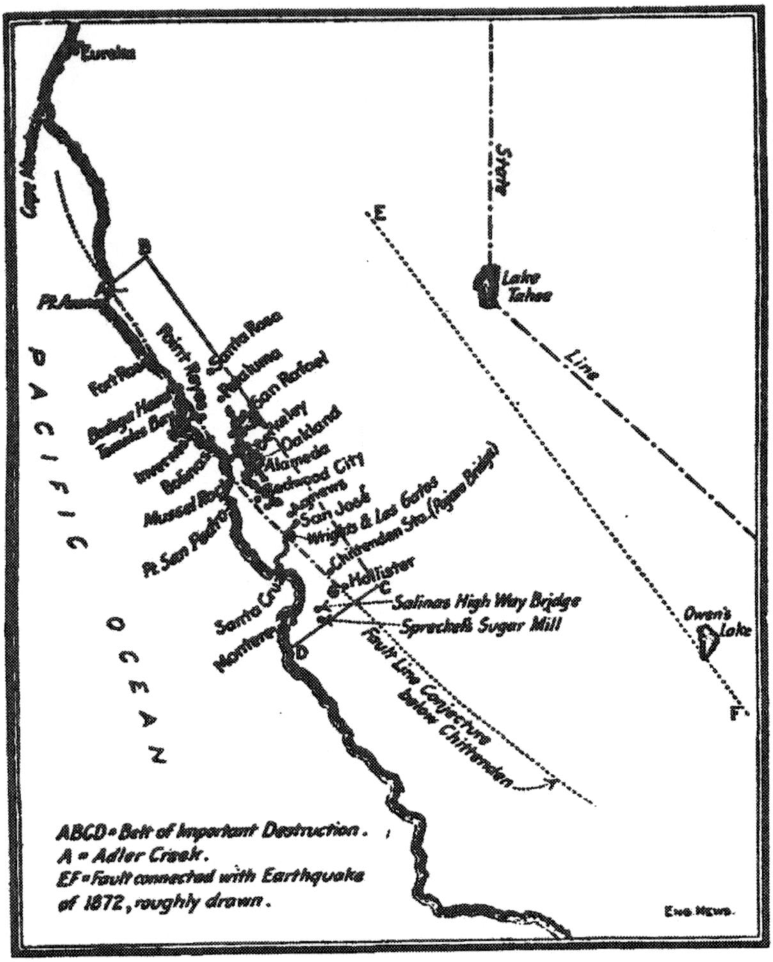

FIG. 2—*Map of California, Showing Position of Fault Line, Earthquake of 1906. Reprinted from "Engineering News," June 28, 1906.*

The Main Coast Range Fault

The earthquake of April eighteenth has affected the crust and the surface of the ground along the main coast range fault. This fault or rift, Fig. 2, runs in an almost exact right line from Point Arena in the north, following along the coast through structural valleys or bays in a south-south-easterly direction, to Hollister in the south. Above Point Arena it disappears into the ocean, although there is some evidence that it curves to the east and approaches the land in the region of Cape Mendocino. This is probably the case; at any rate it would explain the considerable shock felt in Humboldt County. To the south of Hollister the fault may be traced by the observing eye along the western side of the San Joaquin Valley and into the desert lands to the south; some say almost to the head of the Gulf of California. As to the exact location and limit to the fault line in the south I can not speak definitely. I have not examined that region nor does the question of southerly extent really concern the engineer in this present earthquake problem. I have followed the rift from Point Arena to the region of Hollister and Tres Pinos. In this length of a little more than 200 miles the ground along the fault line was broken, and considerable horizontal and some vertical movement occurred throughout that distance. Movement and faulting below Hollister, if there was any, was slight, and I have heard of no authentic reports. Below Hollister, moreover, there is little improved property and therefore hardly anything of engineering importance to destroy. Along the fault from Point Arena to Hollister the earthquake destruction was most severe. Water pipes, conduits, bridges, fences, roads, water courses, all things that crossed the line were crippled or rent asunder. In the north, trees were uprooted, broken and cracked, and everywhere along the fault buildings of weak construction were violently thrown down.

The Belt of Main Earthquake Disturbance

The belt of great disturbance may be said to extend somewhat to the north of Point Arena into Humboldt County, and perhaps somewhat to the south of Hollister and Mt. Pinos, a distance of 300 miles or more. The disturbance was keenly felt for considerable distance to the east and west of the fault line, and a width of fifty miles may be assigned to the belt of great destruction. Within this belt will be found all the important examples of earthquake destruction to engineering works. It is plain that the extent of the seisma is large.

Effect of Topography and Geological Structure

Within this belt, as I have already stated, the most severe racking of structures occurred right on the fault line, the most striking examples being the Pilarcitos conduit and San Andreas dam of the Spring Valley Water Company of San Francisco, and the Pájaro River bridge on the Southern Pacific Railroad near Chittenden Station, where the railroad crosses the Pájaro River. These structures, though well designed, were much racked or even ruined, as in the case of the Pilarcitos conduit, because of the large and unequal movements of the ground along the rift. The intensity of shock, however, is observed to have varied greatly, and well-built structures which rested on hard and rocky surfaces were relatively little damaged. Rocks and the structures resting upon them were shaken by an elastic vibration without differential movement, and whenever the construction was intelligent and honest it withstood the shock. There are many examples of buildings of such behavior on rocky summits in San Francisco. Comparatively little destruction was meted out in cities like Santa Cruz, San Rafael and Berkeley, which rest on rocky foundation, or other sound coherent materials.

Within the belt of great destruction are found many localities or spots where considerable differential surface movement occurred, though at considerable distances to either the east or the west side of the rift. Examination shows such localities to be overlaid with loose incoherent material. In one place we find a loose river deposit, in another a marsh, and in a third an artificial fill. Examples of destruction in such localities are found (1), near Salinas, where the Spreckels Sugar Mill, a first-class steel cage constructed building, was much racked; (2), in the San Bruno marsh, near South San Francisco, where the Crystal Springs conduit of the Spring Valley Water Company was smashed; and (3), on the filled ground areas of San Francisco in the Mission District and along the water front near the Ferry house, where the street surfaces were deformed into billow-like waves, and structures with weak foundations were generally destroyed. In all these cases the differential motions were of a secondary nature and not directly connected with the movements along the fault line. The vibration of the earth's crust caused marshes which were near to the center of the disturbance to shake like bowls of jelly, and loose sandy and alluvial deposits and artificial dumps and fills were much shifted and shaken about upon the firmer ground beneath them.

A line of considerable but lesser intensity of shock may be traced approximately parallel to the fault line through the bottom lands of the valleys which contain Santa Rosa and Healdsburg in the north, and Agnews and San José in the south, with the Bay of San Francisco as a central feature.

These bottom lands are alluvial deposits of soft and considerably incoherent materials. Structures resting upon these bottom lands were severely shaken, though in general there was little if any differential movement of the ground in such cities as Santa Rosa and San José. Structures located nearer the foothills of these same valleys, resting on firmer ground, were observed to suffer very much less.

Within the confines of the city of San Francisco one finds evidence of great variation in shock closely related to and to be explained by the nature of the surface topography. It is a general observation that the earthquake waves transmitted by the softer and less coherent materials and formations appeared to be much more destructive than waves which traversed the hard and more elastic rocks and other sound deposits. The billow-like effects that appeared in the streets of San Francisco near the Ferry house are most excellent examples of deformations in soft, incoherent materials. The sliding and rolling effects observed on some of the sand dunes and especially along the hillside at the northern end of Van Ness Avenue may be cited as allied phenomena. The great contortion of sandy deposits on the south bank of the Salinas River in the vicinity of Salinas and Spreckels is another good example.

Possibility of Future Earthquakes Near San Francisco

I believe that the crust of the earth acts like a more or less brittle skin on the surface of a plastic globe. I believe the interior of the globe to be potentially plastic despite its relatively high density, because of the great pressure to which the material is subjected due to the enormous loads of superincumbent materials. The crust can not be self-supporting like a spherical shell nor locally as a segment of a spherical dome because of the large radius of curvature of the earth's surface and small depth of shell or arched ring. For the earth's crust to act as a self-supporting stable arch or dome would demand the existence of arch ring or dome stresses in the crustal rocks far in excess of the crushing strength of granite. Wherever the pressure from within against the crust is relieved, the crust must sink, and where for some reason the interior increases its pressure against the crust the land must rise. The earth's crust may be conceived to rest like a brittle slab upon an interior of a semi-plastic nature; whenever the conditions of pressure between the crust and the interior become disturbed, the crust must give and adjust itself to put the stresses in the rocks into equilibrium. To produce this equilibrium the crust must give at its weakest point. In this way a crack or slit, or as it is termed in geology, a fault, is produced. Within the confines of California one finds a region of structural weakness, and as has already been pointed out, the State is marked by a number of long fault lines running along the foothills of the high mountain ranges in the Sierra region and along the structural valleys of the coast ranges. Slippings and adjustments of the crust have occurred along these fault lines many times in the remote past, and the present evidences of geological faults and rifts are the accumulations of many past earthquake breaks. When a slipping has once occurred along one of these pronounced lines of weakness, either due to an actual rupture of the rocks or to the sundering of an old break, it is fair to presume that the crust in that vicinity has been set at equilibrium. It is also to be expected that a number of very minor shocks should follow in quick succession after a heavy earthquake. They represent secondary slippings and local readjustments after the main movement. A long period of time must then elapse before a sufficient accumulation of stress can result in the same region to produce another rupture and renewal of movement by overcoming the friction and partial cohesion of an old break. It is probable therefore that a heavy earthquake in the region of the main coast range fault will not occur in the immediate future, and that the crust in the region of San Francisco has been put into equilibrium for a considerable period of time. I believe it probable that Western California will not be subject to a heavy earthquake for at least a century, but earthquakes can not be predicted and another one might come tomorrow. A shock of considerable severity might occur at any time, and it is plain that thinking men must be ready to expect surface disturbances somewhere in California in the next generation.

It is our duty to anticipate these disturbances. Any one who has carefully studied earthquake destruction can not fail to appreciate that great structural losses are due primarily, except in the immediate region of a fault line or upon loose deposits, to faulty design, poor workmanship, and bad materials; let us hope through ignorance and a blind disregard for earthquake possibilities; yet I regret to add that I feel convinced that much of the bad work is due to a combination of criminal carelessness, vicious and cheap construction. Rather than try to tell outsiders that San Francisco was visited by a conflagration I believe that it will do San Francisco and California in general more lasting good to admit that there was an earthquake, and that with honest and intelligent construction and the avoidance of geologically weak locations for important structures, our losses within the earthquake belt would not have been so great.

Destructive Path of the Fault Line

Referring again to Fig. 2, we note that the rift disappears in the ocean at the extreme north a few miles above Point Arena at the mouth of Adler Creek. From Adler Creek the fault may be followed southward approximately parallel to the coast line to Fort Ross, where it runs into the ocean about 2½ miles south of the fort. Earthquake vibrations were very severe in the neighborhood of Fort Ross, I believe more severe than in the vicinity of San

Francisco, but there were no important human structures to demolish. Fig. 3 shows a redwood tree about six feet in diameter which happened to stand right on the fault line near Fort Ross. It was split into halves for a distance of 35 feet upward from the ground. The westerly half was sheared toward the north and actually moved past the east half a distance of about 8 inches. Fig. 4 shows a pine tree which was situated a few feet to the east of the fault line. It was thrown so that it leans toward the east. It was somewhat cracked at the base and its roots on the west side were torn, due to the shearing action along the fault line. The tree was subjected to torsion because it had roots on both sides of the fault, which explains the cracking at the base of its trunk. Many trees were more or less ruptured along the fault where the line traverses timber growths; and in one place I observed leaning redwood trees for a considerable distance distinctly marking the passage of the fault through a forest. Leaning trees make a very uncommon and inspiring sight in a redwood growth, where the trees are over 200 feet in height. At Fort Ross, near the fault, I noticed many partly decayed trees and trees weak in parts demolished and snapped off by the shock, when sound ones close at hand were undisturbed. I could not help thinking how Nature in this way pointed out her weak and her strong construction. In the cities that I have visited within the earthquake belt, from Santa Rosa to San José, one can pick out the black from the white sheep in buildings. Natural and human works behave alike. They are governed by the same mechanical principles.

The fault line is lost in the ocean for some miles south of Fort Ross. It is again noticed as one proceeds southward, where it crosses the sandy spit of Bodega Head, which extends into the ocean. At this point the surface effects are very indistinct and of little moment to the engineer.

Again, the fault line disappears in our southward journey, but it distinctly follows through the structural trough which forms Tomales Bay and appears on land again at the southerly end of that bay near Point Reyes and Olema. From Point Reyes to Bolinas Lagoon the fault line is extremely distinct and this region offers to

FIG. 3—*Redwood Tree, Situated on the Fault Line, near Fort Ross, Sonoma County, California.*

the geologist and engineer equally interesting evidence. At Bolinas the fault disappears into the ocean outside of Golden Gate and does not appear on land again until we reach Mussel Rock on the San Francisco peninsula. Along the coast above Mussel Rock to Lake Merced great coastal disturbances were produced, due to landslides on the steep banks caused by the nearness of the shore line to the fault. Heavy landslides along the coast occurred also to the south of Mussel Rock at San Pedro Point and Devil's Slide, where the preliminary grading for the Ocean Shore Road, which is to connect San Francisco and Santa Cruz, was entirely wrecked.

FIG. 4—*Pine Tree Standing on the Line of Fault near Fort Ross, Sonoma County; the Ruptured Surface Shows the Characteristic Appearance of Newly Plowed Ground.*

From Mussel Rock the fault proceeds southward along a chain of small lakes between the San Bruno marsh on the east and the coast ranges on the west. These lakes are numerous, and the observing eye at once notices a connection between them and the characteristic geological formation which marks the line of the fault. Small basins or ponds, some brackish, some even salt, are of frequent occurrence along the fault line from Fort Ross to Hollister. Continuing southward from Mussel Rock and the chain of small lakes, the fault runs through a long and narrow structural valley, passes along the east bank of San Andreas Lake, follows the valley below that lake, and coincides with the longer diameter of Crystal Springs reservoir.

It then continues southward somewhat to the west of Redwood City and Palo Alto (seven miles distant) and crosses the Narrow Gauge Railroad in the neighborhood of Wrights and Los Gatos. Here mountain tunnels on the railroad have been made impassable by slides so that trains can not cross through the Santa Cruz Mountains from San José to Santa Cruz. Agnews and San José are twelve and thirteen miles respectively to the east of the rift.

The fault may then be clearly followed to the region of Chittenden Station on the Southern Pacific Railroad, where it crosses the Pájaro River about 1,500 feet south of the station and passes right through the Pájaro River Railroad bridge. The fault line then proceeds inward to San Juan Bautista, or farther, where it leaves the region of important structures.

Differential Horizontal Displacement

From Point Arena to San Juan all evidence clearly indicates that the ground on the west of the fault moved north from seven to nine feet relatively to the ground on the east. Straight fences that crossed the rift were invariably sheared so that they are out of alignment or offset from six to fifteen feet. All things that crossed the fault (water pipes, houses, dams, and water courses) were sheared. It is clear to the observer that the ground on the west of the fault moved; that on the east did so too. Probably the material along the fault moved in opposite directions on the two sides, with the resulting displacements mentioned. The tearing of the surface along the fault also clearly shows that there was some torsion in a clockwise direction when the eye looks downward. Fig. 5 shows a fence one-half mile south of Fort Ross which did not collapse, but was curved and bent to suit the lateral movements of the earth, so that its unmoved parts are now nine feet out of alignment. A second fence one-half mile farther south completely collapsed for a few feet on each side of the fault and was offset fifteen feet six inches. Near the head of San Andreas Lake a fence, Fig. 6, was offset seven feet. A roadway in the Point Reyes locality was dislocated about twenty feet. In all cases the west side moved north.

FIG. 5—*Fence near Fort Ross, Sonoma County, California, Offset Nine Feet at the Line of Fault.*

The amount of offset along the fault at the surface is affected by the nature of the surface material. On marshy and on sandy ground and on steep hill slopes and alluvial valley lands, the offset is sometimes more and sometimes less than the average, due to secondary motions of the looser surface materials or to local landslides on the steep hillsides. It would seem that there was somewhat more, perhaps two feet, sliding motion along the fault line at its northern end near Fort Ross than at its southerly end near Pájaro, where the movement diminishes.

Differential Vertical Displacement

The relative horizontal movements along the fault were much more marked than the differential vertical displacements, although the old scarps now somewhat rounded by the weather show considerable vertical differential motion for past times. In the north, in Sonoma County, due to the present earthquake, one observes near Fort Ross a general lifting of the surface to the west of the fault line. The movement does not exceed four feet. In the south there is little, if any, differential vertical displacement.

FIG. 6—*Fence near the North End of San Andreas Reservoir, San Francisco Peninsula, Offset Seven Feet at the Line of Fault.*

Absolute Movement of the Crust

Throughout the disturbed belt there is no doubt that the crust has been profoundly shaken. Latitudes and longitudes have doubtless been shifted a few feet, but it would be difficult to substantiate fully this statement. There is equally little evidence of change in elevation; yet mountain tops have probably been moved in elevation by small amounts. Only careful surveying and leveling and a comparison to geodetic records can throw light upon this question. The triangulation of the San Francisco Bay region is now being checked for this purpose by engineers of the United States Geodetic Survey.

Santa Rosa

Santa Rosa, about fifty-two miles north of San Francisco on the California and Northwestern Railway, is nearly twenty miles east of the fault line. Nevertheless it was visited by great earthquake and fire destruction. Eastern people have heard little of the losses of Santa Rosa because they were overshadowed by the largeness of the destruction at San Francisco; yet in my judgment, proportionately speaking, Santa Rosa's loss was greater than that of San Francisco. The city stands on alluvial ground. Its business center was wiped out by fire and practically every brick building in the business district collapsed in the earthquake. I believe all but two or three of Santa Rosa's brick buildings were razed with the ground by the temblor. But it is my judgment that the shock was less serious in the northern city than in San Francisco.

How then should the general destruction be explained? The brick buildings of Santa Rosa were carelessly constructed. Lime mortar was almost invariably used with bad brick bond. The sand of the lime mortar used was what is locally known as "drift" sand, containing, according to a local engineer, practically fifty per cent of loam. When we remember that it has been too common a custom in California to lay brick for small structures without sufficiently wetting them, in fact almost dry, what else should be expected, especially when we further observe a most inadequate anchoring of the floor and roof frames to the outer walls, and a usual absence of necessary cross walls or frames. Besides brick buildings are not capable of withstanding heavy earthquake vibration. It is a fault of design more than of workmanship. Upon this point I will speak more in detail later in reference to brick structures in the city of San Francisco.

Wooden buildings in Santa Rosa with few exceptions were unaffected by the earthquake except for the general throwing down of chimneys. The few frame houses which were wrecked collapsed because of rotten or faulty and improperly braced underpinning to the first floor or cellar. Frame houses, well built, are adequate for earthquake countries.

No railway or highway bridges in the vicinity of Santa Rosa suffered. No breaks occurred in the water works system, its artesian sources of supply, or its pumps. A few house-service pipes broke at the house lines. No sewers were crippled. It is plain to me that there was little or no unequal resultant movement of the ground at Santa Rosa such as one finds at the fault line. The destruction was due merely to earth vibration and the brick buildings almost without exception collapsed like so many sand piles.

The business portion of Santa Rosa burned because fires started simultaneously in the rubbish and the facilities for fighting fire were too limited to cope with a conflagration. There was never a lack of water, in fact there was more water than under normal conditions, because the people were not using their natural supply; and for the same reason the pressure in the mains was greater than under normal conditions, although some local Santa Rosans ascribed the higher pressure to the effect of the earthquake on the subterranean grounds surrounding the artesian wells.

Fig. 7—*Santa Rosa Flour Mill. A Typical Brick Structure with Wooden Interior, Three Stories in Height; the Major Portion Completely Collapsed.*

Fig. 8—*Carnegie Library, Santa Rosa. A Typical Brick Building with Wooden Interior Framing; the Outer Walls Faced with Cut Stone. Buildings of This Type of Construction Invariably Were Shattered near the Roof Lines. The Picture Is Representative of the Behavior of Stone-faced Buildings with Inadequate Wood Framing.*

FIG. 9—*A Scene in Santa Rosa Photographed Shortly after the Earthquake and Fire. The View Shows the Corner of Fourth and Mendocino Streets. The Ruins of the Court-House face Mendocino Street. On the Left in the Picture Is Seen the Collapsed Keegan-Brush Building, a Two-Story Brick and Stone Structure, Devoted to General Store Purposes. This Building Did Not Burn, and Ninety Per Cent of the Stock Was Saved. On the Right Side of Mendocino Street the Fire Destroyed Everything. In This Respect the Picture Is Instructive in That It Shows the Difference in the Destruction by Earthquake and Fire.*

FIG. 10—*The Collapsed Court-House Dome, Santa Rosa. The View Is Taken with the Camera on the Second-Story Roof; See Fig. 9. The Court-House Consisted of Three Stories of Brick, the Third Story Smaller in Plan Than the Other Two. The Third Story Supported the Wooden Dome.*

Figures 7,8,9 and 10 show typical views of earthquake and fire destruction in Santa Rosa.

San Francisco

To the student of structures San Francisco in the month following the earthquake offered a field of observation so large that I hardly know where to begin to describe my impressions. Within the city limits one found most varied examples of surface and foundation materials, from hard rock upon the hillsides to treacherous, filled ground along the water front of the bay and upon the old stream beds of the Mission Creek. Every degree of construction in building was presented, from the first class steel cage constructed buildings, locally called "Class A," to the cheapest types of brick and frame houses. All grades of workmanship, good and bad; all types of design, scientific and unintelligent; all degrees of construction, from honest to dishonest examples, were to be seen on every hand. On adjoining lots one found the so-called fire-proof structure and miserable fire traps huddled together. Or one saw the ruins of a building carefully fire proofed within, but entirely lacking in exterior protection to resist a conflagration from without. Many of the Class B structures were clothed with iron shutters and gave some evidence that the designer and builder had at least thought of exterior fire-proofing; but within, the building was ready to burn like the contents of a furnace.

Then as one walked through the desolate fire-stricken streets, one was constantly forced to compare the ruins of municipal and private buildings. Government buildings in general were well built, and it is not intended that they should be included in this criticism. There evidently was a difference in the construction of city buildings and those erected by private parties. Certainly there was a difference in their earthquake and fire resisting powers. School buildings and churches too, too often exhibited pronounced weakness in construction.

I believe these statements do not apply to San Francisco only. The attitude toward careless construction for buildings of community interest is entirely too prevalent throughout the United States. The facts have not been brought home in other large cities simply because those cities have not been so sorely tested as San Francisco was in April, 1906. It is perhaps right to say in this connection that it is unnecessary to consider the personal integrity of builders, or to insinuate criticism concerning the ability of the architect. What should be emphasized first and foremost is a general principle applicable to cities and communities all over the United States, namely,—that the public in the building of municipal structures, school houses and churches, expects far too much for the money appropriated. The result therefore is a building of improper construction, and we need not wonder that such buildings showed themselves seriously weak when tested by a severe earthquake, and helpless in a conflagration.

The discussion for San Francisco naturally divides into three main parts:—(1), earthquake effect upon structures; (2), fire-proofing of buildings, and fire-resisting qualities of materials; (3), a critical digest to determine the best materials to be used and the most favorable types of design to be employed to resist earthquake stresses and retard the progress of fire. In this paper I will restrict my remarks almost entirely to the earthquake effects.

Earthquake Effects in San Francisco

As pointed out earlier in this paper, the intensity of shock was not found to be constant over any given area, but was greatly affected by the nature of the ground surface. Structures resting on the rocky hill slopes of San Francisco suffered least. In the swales between the hills, upon comparatively firm ground deposited there slowly by natural processes, one found increased destruction. The shock was felt with still greater violence upon the sand dunes, while the worst destruction in the city was meted out on the artificially made lands near the water front and upon the old swamps. Roughly speaking, then, one may emphasize four varieties of ground:—(1), rocky hill slopes; (2), valleys between the spurs of the hills; (3), sand dunes; and (4), filled ground; upon which, in the order named, was found earthquake destruction of increasing severity.

FIG. 11—*Street Surface in Front of the Ferry Tower, Showing Undulations and Cracks in the Asphalt Pavement.*

But this classification can only be helpful for a preliminary and superficial discussion. Well-built structures on proper, deep foundations stood the shock on soft ground, while buildings of faulty design went to pieces on much more favorable locations. One may be easily led into error in judging of the degree of shock by the amount of destruction. A complete knowledge of the type of structure, grade of workmanship, and properties of materials used, together with the geological considerations, is necessary to establish an intelligent conclusion.

FIG. 12—*Rupture of Car Tracks and Pavement on East Street, Corner of Pacific Street.*

Street and Surface Deformations

Great distortion of the surface was best observed in the streets, and was found on the filled areas and in some places, on the sand dunes. The best localities for observation were:—(1), Market Street near the Ferry Building, Fig. 11; (2), the water front on both sides of the Ferry Building, Fig. 12; (3), the corner of Howard and Spear Streets, where the J. A. Folger Company's building was saved from fire; (4), the corner of Mission and Seventh Streets, the location of the General Postoffice ; (5), Van Ness Avenue at Eddy Street; (6), the north end of Van Ness Avenue and the streets on the hillside slope in that vicinity downward to the water front on the north; (7), Howard Street, between 17th and 18th Streets, Fig. 13; (8), Valencia Street, between 18th and 19th Streets, where the Valencia Hotel was wrecked with great loss of life, and the main water pipes were sheared; (9), Fourteenth Street, between Mission and Howard Streets; and, (10), the water front near the Potrero District. I might enlarge this list, but these are the typical examples. Examples 5 and

6 represent surface distortions on sand dunes. The rest are examples of filled ground deformations. Upon the filled ground the surface was very generally thrown into billow-like waves, a type of disturbance which was best seen near the Ferry house. Upon the sand dunes the surface was shifted by sliding motion so that cracks and fissures appeared upon the streets at right angles to the direction of sliding.

FIG. 13—*Scene Corner of Howard and Seventeenth Streets Showing Rupture of Car Tracks, Sewer, Water and Gas Pipes.*

It was in these areas that the sewers and water pipes of the gridiron system were so generally crushed and broken. Even had the main conduits survived, water could not have reached the hydrants in the lower Mission District. The brick sewers were uniformly helpless to resist destruction in these regions and the cast iron water and gas pipes fared no better. I believe reinforced concrete sewers in these districts would have shown much greater resisting qualities, but I am convinced that even such material could not with-stand earthquake stresses on the dividing line between made and filled ground. At such points flexible joints might have helped the sewer and water pipes, but it is difficult to conceive of a practical means for procuring flexibility at any given point in a brick or concrete sewer. Moreover, it would be requiring a power of prediction not resting in human beings to determine the proper locations for flexibility. Important water pipes wherever possible should avoid soft ground by going around it, and those that must traverse filled areas should be of riveted steel or wrought iron with flexible joints at intervals. Greater probability of resistance to rupture might be further insured by encasing pipes traversing the most treacherous

FIG. 14—Collapsed Frame Houses on Howard Street, between 17th and 18th Streets. The One on the Left Is Completely Razed. At This Place the Earth Movements Were Especially Severe, and Even Good Construction Would Have Suffered.

ground in tunnels of reinforced concrete with suitable clearance between the pipe and the tunnel walls. An added advantage of this scheme would be ready means of inspection.

In discussing earthquake effects upon sewers and water pipes in the filled ground regions, one of necessity is led to consider fire problems as well because the water supply and fire questions are so vitally related. Water mains should be generously provided with a system of cutoff valves commanding the direction of flow through the pipes of any given locality. At the time of the earthquake San Francisco was in dire need of a salt-water system, and immediate steps should be taken to provide such a system and to secure a number of powerful fire boats.

Ordinary Frame Houses

Many frame buildings collapsed on the filled ground, Fig. 14; and an unreasonably large number on the more substantial areas of the city. Some of the destruction in the made land district was unavoidable, but much of the general collapse of frame houses was due to improper underpinning in the foundations. Such houses virtually stood on stilts; consult Figs. 15 and 16. Moreover, the cheaper frame buildings were provided with carelessly constructed brick chimneys built with weak lime mortar; naturally the brick work cracked to pieces. It is therefore easy to see why so many of these houses burst into flames immediately after the earthquake. It is said that fifty-seven fire alarms were sent to the Fire Department within the first half hour. Of course not all of these fires were produced by fallen or cracked chimneys in frame buildings. Some were similarly started in cheap brick structures, which are just as readily inflammable,—and still others by the breaking of electric wires of high tension, or through improper electric insulation of power wires.

Carefully built wooden-framed houses are especially well adapted to withstand earthquake shock and represent a most desirable type for small dwellings. The weak feature is the brick chimney. But even this can be built so that cracking will not be so complete as to cause fire. For earthquake countries wooden buildings should be intelligently framed to act like an elastic cage or unit, and the foundations should be carried below the surface sufficiently to

Fig. 15—*Frame House Wrecked in Santa Rosa; Due to Weak and Decayed Underpinning.*

Fig. 16—*Collapsed Frame House in San José Showing the Effect of a Lack of Transverse Framing and an Absence of Continuity in the Vertical Sticks at the Floor Levels.*

reach firm ground. It is too common a practice which completes one story of a framed house before starting the frame for the next story. Where there is no continuity in the main wall framing from floor to floor, weakness exists at the floor level; consult Fig. 16.

If brick chimneys must be used it is a mistake to use strong mortar above the roof line. With strong mortar a chimney top will fall in one piece and crash through the roof; with weak mortar it will disintegrate and the individual brick will roll off the roof.

Ordinary Brick Buildings

The most general destruction by earthquake in San Francisco was observed in ordinary brick buildings. Brick walls were usually thin, of careless bond, and built with lime mortar of little strength. Apparently many brick walls were laid without wetting the brick before applying the mortar, because one found so many cases where the crumpled piles of brick showed clean surfaces, the mortar not having adhered to the brick to any degree. These buildings were of timber framing within, and the floors and roofs were improperly anchored to the outer walls. The earthquake shook out whole sides of brick buildings because there was no provision for a proper tying or anchoring of the walls from within. The so-called "fire walls" above the main cornice were thrown down in almost every case. Had the earthquake occurred later in the day, the falling of brick from fire walls would have caused many deaths.

Where buildings of this class had trussed roofs of timber, the framing was generally imperfect. These roofs commonly rested on inclined rafters with no bottom cross tie to keep the rafters from spreading. The earthquake vibration tended to drop the peak of the roof and spread the ends, which, in kicking against the walls, forced them out and threw them down. In this way a number of brick school houses were severely shattered. Many brick buildings were built like weak boxes, with no adequate provision for transverse stiffness, because the structures entirely lacked transverse brick walls or frames. Brick structures in San Francisco and vicinity should never be built with lean lime mortar. One part of cement to four or five of lime mortar would give very much better results. At least every other joist of the floor system ought to be carefully tied to the outer brick walls. The roof similarly should be anchored to the walls; and its main trusses should have properly designed lower chord tie rods, and not depend for their resistance to spreading upon the stiffness of the side walls of the building. Foundations of these structures should also be made with great care to have them act as units to prevent unequal settlement.

The prime requisite for a structure to withstand earthquake shock is elasticity; that is, the ability to return without serious damage to its original shape and position after being distorted. It should vibrate without offering great resistance to distortion; in other words, it should yield readily. The wooden framed and the steel framed building answer this requirement. To an almost equal extent the reinforced concrete building does so also. But structures of brick and stone built of blocks, like brick work and cut-stone masonry, with horizontal and vertical mortar joints do not answer the requirements of yielding and elasticity to any desirable degree. This is especially true when the interior framing is of wood and of the weak and faulty design depicted in the preceding paragraphs. Brick and stone walls when not sufficiently reinforced and intelligently strengthened and supported by steel framing are inadequate for important buildings in an earthquake country. Consult Fig. 17.

While I have been decidedly severe in condemning block-work construction I must admit that some brick structures made a good showing. They are the exceptions that prove the rule. For every brick building that withstood the shock, it is easy to give a number of examples of complete failure. St. Patrick's Seminary, at Menlo Park, is an example of most excellent brick work, yet it was decidedly shattered in the main towers. It is not unreasonable to say that where brick structures withstood the shock they did so in spite of the fact that they were built of blocks or because the shock was slight or the building favorably situated.

Fig. 17—*A San José High School. Ordinary Brick Construction with Wooden Interior.*

Fig. 18—*St. Francis Church, San Francisco; an Example of Excellent Brickwork.*

In San Francisco there were a number of large brick structures of most excellent type which weathered the earthquake with practically no damage. I refer to several churches, see Fig. 18, to the Custom House or Appraiser's Building and to the Palace Hotel. The walls of the Appraiser's Building were thirty-six inches thick. The Palace Hotel was built immediately after the earthquake of 1868 and was intended to be earthquake-proof. The outer walls were most carefully designed, the window openings were crowned with arches and the interior of the building was divided into compartments by numerous cross walls running parallel to both sides of the building. Looking down upon the ruins of the structure one saw a honeycomb of brick walls giving lateral stiffness in all directions. Iron rods were embedded in the walls. The old Palace Hotel therefore should not be classed as a simple brick building, for it contained some attempt at reinforcement.

Stone Structures

FIG. 19—*First Baptist Church, Oakland, California. The Dangerous Tower Was Pulled Down Several Days after the Earthquake.*

Heavy stone structures, without steel frames, especially where the outer walls were not properly tied to the interior floor and roof frames, suffered severely throughout the earthquake belt; see Fig. 8. Too many structures of this type for architectural reasons are made top heavy; see Fig. 19. Much of the recent construction at Stanford University may be criticized along these lines. In Oakland, Berkeley and San Francisco, heavy masonry church towers were invariably demolished, except where they were properly reinforced by interior steel frames or were strengthened by interior cross walls and tying rods. Heavy stone ornamentation should be discouraged and heavy stone cornices should be avoided. Where architectural effect is insisted upon, no expense should be spared in anchoring heavy stone cornices by the use of metal.

Chimneys

FIG. 20—*Fire Ruins of a Power-House of the San Francisco Gas and Electric Company, Station C. The Chimney of This Structure Fell During the Earthquake Shock, Demolished a Part of the Equipment and Killed One Person.*

Chimneys in San Francisco were built of brick and very often without cement in the mortar. With few exceptions the chimneys were thrown down by rupture within the middle third of the height. A number of lives were lost by falling power house chimneys; see Fig. 20. In the future I believe chimneys should be built of reinforced concrete and not of brick. Where brick is insisted upon, the bond should be carefully provided for, the mortar should be rich in cement and there should be some metal reinforcement.

Class B Buildings

FIG. 21—*The Mills Building, San Francisco; Self-Supporting Brick Walls, Interior Framing of Steel, Floors and Partitions of Hollow Tile.*

There is little to say regarding the earthquake effect upon the so-called "Class B" buildings. These structures consisted of self-supporting walls with an interior framing, partly of metal and partly of wood, with cast iron columns too often in evidence. For large structures of considerable height, one certainly should not advocate self-supporting walls for earthquake countries. The Mills Building in

FIG. 22—*Fire Ruins of the Cowell Building, a Class B Structure, Showing the Great Destruction by Fire.*

San Francisco, Fig. 21, an otherwise excellent structure of ten stories, was designed with self-supporting walls and with main columns breaking joints at every floor level. After the earthquake the walls of this structure leaned from seven to nine inches into Bush Street. While the structure did not appear to be severely racked by the earthquake, the fact that the walls bore little relation to the interior frame, and the interior frame lacked continuity of columns at every floor, would not lead me to recommend such a type of structure for an earthquake country. Moreover, the type exhibits great weakness in resisting fire because the walls tend to leave the frame. In Class B buildings it was a general observation to note heavy cracks running diagonally in the outer walls. Such X-shaped cracks are also found in the brick work of steel-cage constructed buildings, but they obviously produce much more serious consequences in the Class B building, which has no steel frame upon which to depend for unity and coherence.

Reinforced Concrete Buildings

There were no reinforced concrete buildings in San Francisco because before the fire there had always been successful opposition to their introduction. In a few buildings reinforced floors and columns had been used, but there were no outer walls of reinforced concrete. Reinforced floors were common in Class A structures, where they shared favor with hollow tile floors. The little San Francisco evidence that one finds, considering also a few reinforced structures, or partially reinforced structures in other places, such as in Oakland and Palo Alto, leads one to the conclusion that buildings scientifically designed in reinforced concrete present admirable qualifications for earthquake resistance. There is no reason why reinforced concrete cage constructed buildings of at least six or eight stories in height should not be built in San Francisco. A reinforced concrete structure, when intelligently designed, generously proportioned and honestly built, is a monolith of great coherence and high elasticity, combining the very properties best able to resist earthquake vibration.

Class A Buildings

Class A structures stood the earthquake shock admirably. Their steel frames were not materially impaired. These buildings vibrated like tuning forks. It appears that the base of such a structure oscillated and moved with the earth, while the top tried to remain quiet. This statement is substantiated by the general evidence that less shock was felt in the upper stories of high buildings than in the floor levels near the street. Again it seems to be a fact that books and other loose objects were less disturbed and thrown down in offices in upper stories than in those near the basement. The main steel columns of Class A buildings should be as continuous as possible from roof to cellar. Splices should be generously built, and it is always wise to run column pieces through several floors. These high structures were subjected to a considerable racking stress at floor levels, as is evidenced by the lines of rupture in brick walls at the floor horizons. In the future I believe that steel frames for Class

A buildings should be provided with considerable knee-braced framing in the vertical planes between the main floor girders and columns to which they rivet; and wherever possible, diagonal framing should be introduced similar to that provided to resist wind stresses. Wind stresses very commonly are fictitious or imaginary, and certainly never approach the limits provided for; but earthquake stresses, while they fortunately do not occur very often, are intense and of the nature of impact forces. The engineers upon whom may devolve the design of Class A skeletons in the immediate future should insist upon their structural requirements with greater determination than in the past, and not allow the architect to injure the strength of the building for the purpose of securing some less necessary architectural feature or embellishment.

Earthquake forces, relatively speaking, are unlimited in amount when the strength of human structures is under consideration. The amount of earthquake stress produced in a member of a structural steel frame is directly proportional to the resistance offered. The stiffer a structure, the greater will be the induced stress produced by earthquake vibrations. The more a structure is capable of yielding, like a willow tree to the storm, the less will be the tendency for earthquake rupture or collapse.

A committee of the American Society of Civil Engineers in its report on buildings states: "Sufficient evidence is at hand to warrant the statement that a building designed with a proper system of bracing to withstand wind at a pressure of thirty pounds per square foot will resist safely the stresses caused by a shock of an intensity equal to that of the recent earthquake." For bridges and towers, engineers usually provide diagonal members to resist wind stresses. They would use diagonal framing in high buildings also, except for the objection of the architect. It is my judgment that diagonal framing is not desirable for high buildings in earthquake countries. Such framing consists of triangular parts. Geometry teaches us that a triangle can not change its shape without changing the lengths of its sides. Triangular framing therefore is stiff and unyielding and calls forth earthquake stress to the full capacity of the diagonal wind members. If a stiff frame of triangles is subject to earthquake vibration of severity, rupture at the weakest places is extremely likely. The effect of the earthquake on the Ferry tower in San Francisco justifies the above statement. In this tower diagonal wind bracing was used. In the fifth and sixth floors of the towers the diagonal rods were either buckled or ruptured or broken at the eye-ends. In some cases the rivets connecting the gusset plates and angles sheared off, the stress finding in each case the weakest point. The stresses exceeded the equivalent of a thirty-pound wind pressure many times. This is proven by the permanent set or elongation of the diagonal rods. In five cases, including three two-inch square rods, it was necessary to cut off the screw ends to take up the slack. The total stretch in some cases amounted to more than three inches.

For earthquake conditions triangular framing for high buildings is not so desirable as a rectangular framing with stiff joints and continuous members. In steel buildings, rectangular framing is best produced by substantial continuity in the main columns and by bracing these columns with deep horizontal spandrel girders; or by more shallow spandrel and floor girders strengthened with heavy knee braces.

Unlike the triangle, the rectangle can change its form without changing the lengths of its sides. With spandrel girder and knee bracing, therefore, the main columns by their elasticity and continuity can yield and vibrate to a considerable extent without endangering the integrity of the building's frame.

FIG. 23—*The Monadnock Building, Showing Heavy Diagonal Cracks in the Brick Curtain Walls along the Northeast Corner for the Full Height of the Structure.*

Fig. 24—Flood Building, Northwest Corner, Showing Heavy Earthquake Cracking of the Sandstone Masonry Enveloping the Steel Corner Column.

I therefore conclude that the best type of framing in a steel skyscraper consists in the generous use of deep spandrel girders, preferably of the latticed truss type. The new Humboldt Savings Bank on Market Street is an excellent example. The old Call Building has an excellent lateral framing.

In the basement of the Flood Building, on the south side, a number of rivet heads were sheared off in the connections between the girder beams and column shelf angles upon which the beams rested. These ruptures were undoubtedly produced by racking motion coincident with the earthquake vibration. Racking stress in buildings certainly had its greatest destructive tendency in the first floor above the ground.

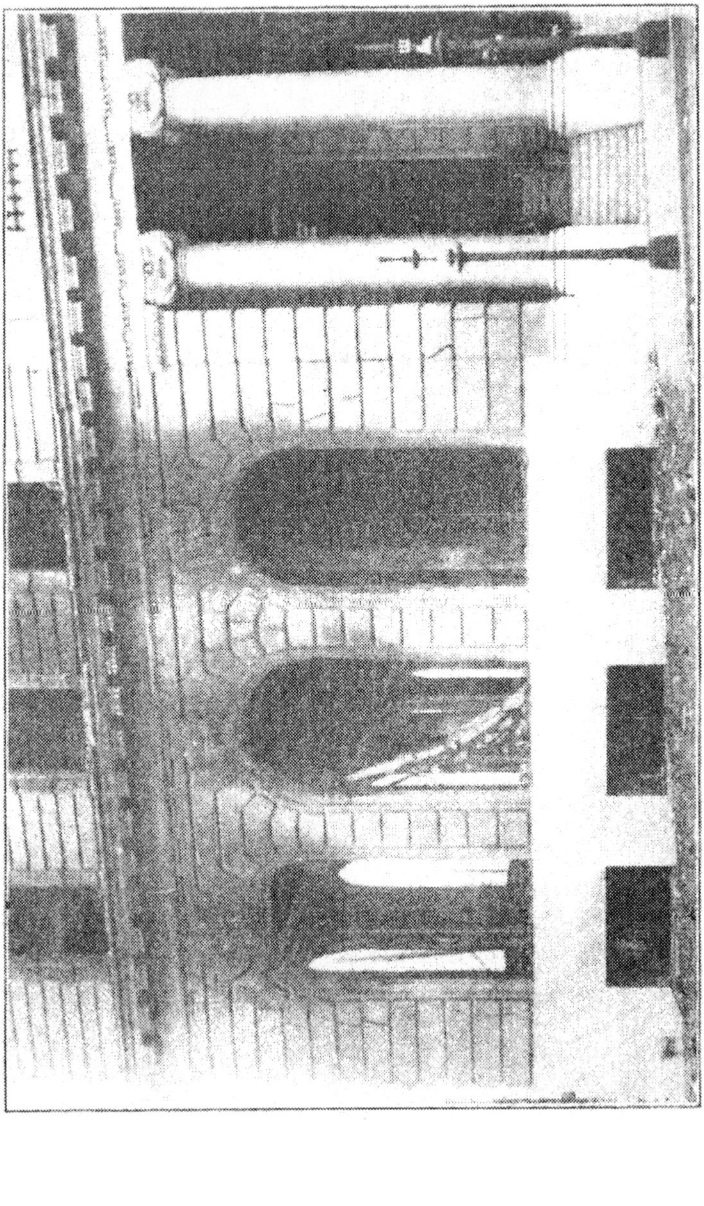

FIG. 25—St. Francis Hotel, Southeast Corner, Showing Heavy Earthquake Cracks in the First Story of the Sandstone Front.

Earthquake vibration in high buildings puts a considerable stress upon the floor connections and columns, but does not tend to produce much destructive effect in the floors; on the other hand, it is clear to see that it does produce a very considerable shearing stress upon partitions, tending to destroy and crack them. This observation leads me to state that I do not advocate hollow tile partitions for high buildings in earthquake localities. Reinforced concrete partitions are much more suitable. Double metal lath and plaster partitions also have merit. I have no particular grievance against hollow tile construction, but partition tile is not a good earthquake material. It is essentially brittle, and difficult to build into a coherent mass. It can not stand flexure or distortion of any kind. For similar reasons I believe there is decidedly more merit in reinforced concrete floors than in hollow tile floors.

FIG. 26—*Fairmont Hotel; North Front, Showing Earthquake Cracks in the Terra-Cotta Veneer of the Second Story Walls.*

The earthquake vibration very generally produced X or so-called "earthquake cracks" in the outer curtain walls of Class A structures. It was very common to see brick work shattered between windows;

and on large side walls, without window spaces, examination showed cracks running at angles of 45° to the horizon. Often heavy cracking was observed at a corner of a building, see Fig. 23. Such destruction is not so serious in Class A buildings as in those designated Class B, which have self-supporting walls, because it is a simple matter to restore the wall at any floor level of a Class A structure without disturbing the rest of the building. Much of the heavy cracking found in Class A buildings may, however, be attributed to improper bond or anchoring in brick work, and to mortar too lean in cement. To me a building with a properly designed steel skeleton and light walls of reinforced concrete, supported by the main frame at each floor level, represents a type containing very much more merit to withstand earthquake shock than structures with brick and terra cotta curtain walls. Such reinforced concrete curtain walls can be made lighter than brick and terra cotta walls, thus reducing the dead weight of the structure. Brick, stone, and terra cotta curtain walls can be safely used, however, especially when carefully anchored to the structural frame.

FIG. 27—*General Postoffice, Southwest Corner, Showing Severe Distortion and Subsidence of the Sidewalk and Street Levels.*

Exposed side walls of a number of Class A buildings had face brick finish on ordinary brick curtain walls to give a dressed appearance. In most of these cases the face brick were tied to the backing by no other means than what is commonly called the clipped course bond. The racking of the earthquake threw out large areas of such face brick. The best example that came to my notice was found on the west wall of the Merchants' Exchange Building. The clipped course bond for facing brick should be prohibited.

I have already stated that in high buildings the earthquake racking was most severe near the street level. One found in nearly every Class A building prominent cracks in the first story walls (Fig. 24), especially at points where heavy weights were concentrated upon corner columns. The stone work of corner columns was severely cracked where the masonry rested upon concrete under the sidewalk, but hardly at all where the stone masonry was carried by the steel frame. Main columns acted like stilts to support the superstructure above the ceiling of the street floor, and prevented the collapse of many a building by column rigidity and continuity at the floor connections in the level of the second floor. It is plain that such floor connections can not be too carefully designed. The most pronounced cracks at first-story columns were found in the Flood Building. One saw them also to good advantage on the sandstone front of the St. Francis Hotel (Fig. 25). Cracking of this character is not an evidence of weakness in the body of the building.

In Class A structures terra cotta was very generously and profusely used for ornamentation on the facing of the curtain walls of the street fronts. Being a material of the same nature as hollow tile, it is brittle and lacks toughness. It was generously spalled by earthquake vibration, and, while I by no means wish to imply that its use should be prohibited, I think it should be very carefully selected, be well burned, be used with greater thickness than in the past, and the hollows in the blocks should be filled. Highly ornamented terra cotta should be discouraged. The earthquake produced considerable spalling in the terra cotta of the Mills Building, and very generally damaged the terra cotta face of the Fairmont Hotel, especially on the north front (Fig. 26).

The foundations of high Class A buildings with small bases can not be designed with too great care. Where such structures rest on sand or near the filled ground areas, the foundations should be deep. Deep pile foundations have given excellent results, even on the filled ground. It is a general observation that structures resting on pile foundations in the made ground were much less affected than adjoining buildings on shallow foundations. The cable roadway on Market Street, where it approaches the Ferry Building, was founded on piles, and it sank little when compared to the general dropping of the street surface along the sides of the tracks. Other types of foundations, such as that under the Call Building, which is of the solid slab variety, have given equally good results.

Intelligently built foundations, not only for buildings but for all kinds of structures, were unaffected by the earthquake. Well-built foundations are adequate. Depth is a prime consideration.

Important structures should not be built on treacherous ground. The site of the general Post office (Fig. 27) was very unhappily chosen. That building, while exceedingly well built, rests on filled ground, and under its southwest end ran an arm of the Mission Creek. In this vicinity the ground was very much disturbed, and the street fell away from the building. The structure, while it stood and was saved from fire, was severely racked, and its beautiful granite fronts were badly cracked on all sides. The interior partitions of the building were of double hollow tile, and were greatly destroyed. The structure represents an example of a building with heavy masonry walls with a partial and too light a steel frame. Considering the material upon which it rests, the building withstood the earthquake surprisingly well.

Fig. 28—Earthquake Cracks on Sandstone Piers of the Ferry Building. These Piers Enclose Steel Columns with a Complete Air Space between the Steel Columns and the Stone Blocks. This Explains the Large Amount of Cracking and Shows That the Stone Piers Are Not Required for the Support of the Second Story.

The Ferry house exhibited another example of excellent construction in a treacherous location. Due to the movements of the filled ground, there was no doubt a considerable pressure brought to bear against the structure from the land side, tending to thrust it into the bay. The building rests on piles in large clusters, and withstood its severe test admirably. The top of the south end wall fell out. The stone work near the street level on the main pilasters was badly cracked (Fig. 28). In the second story horizontal joints were opened throughout the west front length of the building. The marble plates on the interior walls were much broken. The tower was shaken like a tree or flagpole in a vertical plane east and west, nearly coincident with Market Street's length. The east and west faces of the tower were therefore badly racked, just above the roof of the main building, and the stone facing at these places was thrown down, causing considerable destruction in the interior in the central portion of the building below the tower, see Fig. 29. The upper portion of the tower was originally designed in stone, but the consulting engineers advised the substitution of metal sheeting to reduce the weight, and this fortunately was done. Had the upper portion of the tower been of stone, I believe the destruction would have been very much more serious. The stone facing of the tower was backed with brick, and this masonry was supported by a steel frame, as in Class A buildings. It appears that the steel frame was built too light, especially the diagonal members. These diagonals should have been heavier, and of stiff sections instead of rods. Some of the diagonal rods were snapped and some gusset plate connections and riveted joints were ruptured. I have already referred to the superior merit of quadrilateral framing, using spandrel girders and knee bracing. The tower was rapidly dismantled, the steel frame strengthened and repaired and reclothed with reinforced concrete curtain walls.

FIG. 29—*The Ferry Tower Shortly after the Earthquake.*

The Fire Damage in San Francisco

The lessons to be learned from the San Francisco conflagration are many, though not new. A discussion of this subject will in the main repeat the results of the studies of the Baltimore fire. Nevertheless, the question of fireproofing in San Francisco should and will receive the closest attention. It is a very large and important subject and of sufficient dignity to warrant a description by itself. Being without the purpose of this paper, it is not here treated. I will only add that such a discussion should pay some attention also to the fire calamities in Santa Rosa.

Earthquake Destruction to the San Francisco Water Works

The important destruction to the works of the Spring Valley Water Company occurred between San Mateo and the city of San Francisco. There was practically no damage to the water sources and works in Alameda County.

The Alameda conduit between the source and Burlingame lies in a region of lesser disturbance, and appears to have been subject to about the same degree of shock as the Contra Costa Company's works, which supply water to Alameda, Oakland and Berkeley. It is significant to observe that there seems to have been little or no damage to the submerged part of the Alameda conduit where it runs under and across the southern end of San Francisco Bay.

The map, Fig. 30, clearly shows the water works property of the San Francisco Bay cities within the region of considerable earthquake disturbance. This map is taken from the Earthquake Reports of San Francisco members of the American Society of Civil Engineers; consult the Proceedings of that Society for March, 1907.

The Pilarcitos reservoir is to the west of the main fault line, and is separated therefrom by a range of hills known as the Sawyer Ridge.

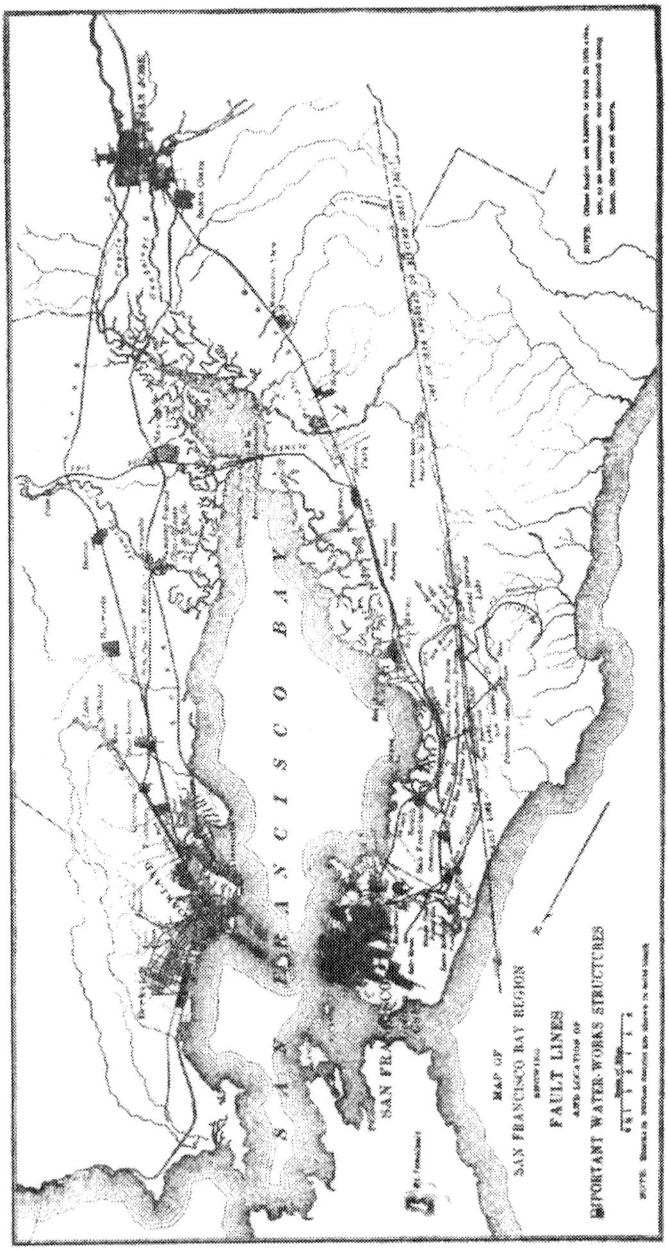

FIG. 30—*Taken from the Report of the Committee on the Effect of the Earthquake on Water-Works Structures, American Society of Civil Engineers; Proceedings, Vol. XXXIII, March, 1907.*

That reservoir is thoroughly intact, and its 95-foot earth dam is unaffected. The waste-way conduit connecting it with San Andreas Lake is also intact.

The main fault line runs through Crystal Springs Lake, but in no way appears to have affected the imperviousness of its bottom. The older Crystal Springs dam, which separates that lake into halves is crossed by the fault. Evidence on the roadway and roadway fences over the dam shows that the dam was sheared about five or six feet in the manner already explained under the heading "Horizontal Differential Motions Along the Fault." This dam, like the San Andreas dam, aside from transverse cracks parallel to the fault line, also exhibits a number of longitudinal cracks along the roadway. When the huge concrete dam was built at Crystal Springs, the older dam became submerged, and to produce the roadway a fill was placed upon the dam of less coherence than that of the material of the original dam. It is impossible to determine whether the imperviousness of this dam below its upper and newer portion has been affected by the shearing action because the water is at the same level on the two sides. From the behavior of the San Andreas dam, however, I am led to believe that the shearing, even of five feet, has not been sufficient to cause serious perviousness, and should the water be released from one side of the dam, I believe it would be retained on the other. If my argument is valid I believe it indicates one advantage of a properly constructed earth dam with a clay core over a light dam of masonry and especially a light arched masonry dam of bold design like the Bear Valley, in Southern California.

The huge concrete dam at Crystal Springs, 115 feet in height above the natural surface, is parallel in length to the fault line. Its curvature is slight compared to the dimensions of its cross section. Arch action is negligible. The dam should be considered as a straight gravity dam. It was probably subjected to a series of thrusts and pulls in vertical planes along its length since it parallels the fault. Its inner face has a much heavier batter than the Rankine or Wegmann calculations would require. The engineer of the dam, Mr. Hermann Schussler, states that he made the batter of the inner face one in four because of earthquake possibilities, he having experienced the earthquake of 1868. This dam is practically unaffected by the earthquake. Some who have examined it state that they have found slight cracks near the base of the downstream toe, but I did not see them, and they certainly are not serious. The intake works at this dam, the Crystal Springs Pumping Station, and all accessory construction in the neighborhood were left practically intact by the earthquake.

The fault touches the eastern edge of San Andreas dam, an excellent construction of earth and clay, 93 feet in height above the original surface of the ground. The fault line is nearly at right angles to the dam. As an eye witness I am convinced that this dam was subjected to a most severe earthquake shock, and since it retains the waters of San Andreas Lake, just as well as before the earthquake, it should be a source of great satisfaction to its designer and builder. Skillfully designed and well built earth dams have been proven by our great earthquake to be structures of great stability, deserving of increased confidence.

At the San Andreas dam the ground on the eastern bank was considerably scarred by cracks running northwest, where the fault line crosses the nose of a hill which naturally projects to form the dam's abutment. There must have been some motion at this point, possibly five or six feet, but it is not clear on the surface. The cracks which were pronounced in this nose or abutment were in the abutment and not in the dam itself. There were a number of smaller cracks running in the same direction at the extreme westerly end of the dam. On the dam's roadway there were small longitudinal cracks to be observed throughout its length, apparently due to the unequal settling of the triangular masses with respect to the core, but they were not serious. I am convinced that an earth dam properly constructed will stand a violent shock. Wherever possible a dam should not cross a geological fault at right angles.

The wooden flume starting near the Crystal Springs Pumping Station, which enters the San Andreas reservoir near the east end of its earth dam, completely collapsed just below that dam where the main fault line crosses the flume. At this place the flume was supported on a wooden trestle about fifty feet in height. In the same locality a waste-way conduit built by day labor, of selected brick with cement mortar, provided a discharge from San Andreas Lake into the valley below the dam to discharge water into Crystal Springs reservoir. The gate-house for this conduit is near the eastern end of San Andreas dam, about 300 feet from the fault line. It is built of selected brick with cement mortar, and exhibits no cracks whatever. About 1,000 feet below the dam the brick conduit curves to the west to its discharge point, and in so doing crosses the fault line. At this crossing the conduit was sheared completely, but due to the excellence of the mortar, the brick was sheared more readily than the cementing material of the joints. This speaks volumes when we reflect upon the general destruction of brick structures throughout the earthquake belt.

FIG. 31—*Rupture of the 44-inch Crystal Springs Conduit on the San Bruno Marsh; Picture Taken May 2, 1906. The Wooden Supports under the Pipe Are Temporary Forms Built after the Catastrophe. The Pipe Has Been Straightened Preparatory to Repairing the Transverse Riveted Joints.*

The Crystal Springs conduit was not damaged between Crystal Springs dam and San Mateo, but between that city and San Francisco it was ruptured in a number of places where it crosses the marshes. It is mainly a 44-inch laminated wrought iron pipe, ¼ inch in thickness, with riveted joints and rivets ½ inch in diameter. The worst destruction occurred in a distance of about 1,600 feet where the pipe crosses a salt marsh between San Bruno and South San Francisco. Here the pipe rested upon a wooden floor supported by pile bents. These piles on the average penetrated the mud to a depth of about 40 feet. The salt marsh evidently shook like a bowl of jelly, the vibration being mainly in a south-southeasterly direction or nearly at right angles to the length of the trestle. It appears that during the vibrations of the earthquake the trestle moved with mother earth. The pipe, due to its inertia, tended to remain quiet. As a result the pipe was alternately thrown from one side to the other of the trestle floor and its wooden box covering was generally smashed. The pipe broke at transverse circular riveted joints, sometimes by tension, and sometimes by crushing. Fig. 31 exhibits a view of one of the ruptures on the San Bruno marsh. Figs. 32 and 33 show other scenes along this pipe line during the period of repair immediately following the earthquake. Redwood planks were used for the temporary sills and stringers because larger timber was not quickly available.

The Pilarcitos conduit for a considerable distance practically coincides with the main fault line. Indeed, one might almost imagine that the break in the ground was purposely staked out along the pipe line, or vice versa, from a point somewhat below San Andreas dam to Frawley gulch, a distance of six miles. In this length the conduit is 30-inch laminated wrought iron pipe, about 3-16 inch in diameter. The center line of the pipe is usually found about three or four feet beneath the ground. In these six miles of length the pipe was ruptured in a great many places, at one place by tension and at another by compression. The direction in which the pipe line crosses the fault determined whether the pipe was torn apart or telescoped. Nineteen ruptures were observed by me from a point near the northern end of San Andreas Lake to Frawley gulch, a distance of about three miles. All ruptures occurred at transverse

FIG. 32—*Rupture on Crystal Springs Pipe Line, San Bruno Marsh.*

FIG. 33—*Crystal Springs Conduit between San Mateo and Millbrae. Temporary Cut-Off to Stop the Flow of Water toward San Francisco and Yet Maintain a Pressure and Supply for the City of San Mateo.*

riveted joints. There were some places where the pipe collapsed; in one instance, for a length of about fifty feet. There were no doubt many more ruptures in this length but they had not been uncovered. At tensile breaks the pipe was pulled apart by amounts varying from almost nothing to as much as five or six feet. At compensating places the pipe was telescoped by similar amounts. Fig. 34 shows a break in this pipe line near the up-stream end of San Andreas Lake. The pipe at this point was pulled apart 53½ inches. A property fence, Fig. 6, which crossed the pipe line about ten feet to the south of this rupture, was offset seven feet along the fault line, the two parts of the fence remaining straight and parallel to their original direction. At Frawley gulch the conduit crossed a timber trestle heavily built, about 100 feet in length and some 25 feet in maximum height. Some of the timbers of this trestle were partly decayed, but the structure certainly was not weak. This trestle was about ¼ mile to the east of the fault line. Nevertheless the shock was so severe that it entirely demolished the trestle and pipe which it carried. The trestle probably vibrated in a vertical plane normal to its length, and was thrown down-stream to the southeast, as shown in Fig. 35.

From Crystal Springs reservoir to Lake Merced the surface of the ground usually is what is known as black adobe land. In places it is yellow adobe and sometimes it is a mixture of the two. The series of ruptures in the Pilarcitos pipe line to Frawley gulch inclusive are therefore found upon firm ground unlike the material at the San Bruno marsh. The destruction was due to the proximity of the fault line or actual coincidence of the pipe line therewith, and as has been repeatedly shown by the evidence of this paper, construction no matter how good was unable to withstand the stresses along the fault.

I have not examined the Pilarcitos pipe line beyond Frawley gulch to the city, but ruptures were probably not so numerous nor so serious in this length. The same remark applies to that portion of the Pilarcitos pipe line near Pilarcitos reservoir. The Pilarcitos

Fig. 34—*Rupture on Pilarcitos Pipe Line near North End of San Andreas Reservoir.*

conduit must be abandoned. Figs. 36, 37, 38 and 39 show additional ruptures in the Pilarcitos pipe line between San Andreas Lake and Frawley gulch. In Fig. 36 the pipe line, supported on a timber framing, crosses a small swale. The pipe was telescoped 41 inches, and the blow-off valve connection was partly ripped from the conduit. Fig. 37 shows two ruptures a little to the north of the break in Fig. 36. At the point where the man is standing the fault line crosses the pipe line at an angle of 45°. The pipe was sheared so that its parts are 20 inches out of alignment, and are moved toward

FIG. 35—*Collapsed Trestle at Frawley Gulch, Pilarcitos Pipe Line, Spring Valley Water Company.*

each other 40 inches. Fig. 38 is a closer view of the same rupture. Fig. 39 shows the pipe crossing a small gulch. The blow-off valve situated at this point was stripped from the pipe and thrown ten feet. The pipe was telescoped 49 inches and thrown out of alignment, as shown. This rupture strikingly shows that the pipe metal was of high quality.

The San Andreas pipe line lies between the Pilarcitos pipe and the Crystal Springs conduit; it is therefore not so near the fault line as the Pilarcitos pipe and avoids the marshes in the region of the Crystal Springs conduit. Its injuries were relatively slight. Only one important rupture came to my notice. It was quickly repaired by putting temporary banding on the pipe.

FIG. 36—*Telescoped Rupture, Pilarcitos Pipe Line, Spring Valley Water Company.*

All the main conduits of the Spring Valley Water Company, with the exception of the San Andreas conduit, were considerably damaged, and in the fortnight following the earthquake no water reached the city from the Pilarcitos, the Crystal Springs, or the Alameda sources, although water from Pilarcitos might have found its way through the waste-way conduit to San Andreas Lake, and thence to the city through the San Andreas conduit after its repairs. San Andreas Lake temporarily became the distributing reservoir for the city and probably will so continue in the future.

It is plain that the water works of the city of San Francisco was subjected to a test more severe than the hand of man could devise. The water supply of a city is a most important matter, and ever since the earthquake the water problem has been the greatest one of many problems which have concerned the citizens of San Francisco. It is not surprising therefore that much concern was

Fig. 37—Diagonal Ruptures, Pilarcitos Pipe Line, Spring Valley Water Company.

expressed about the water situation, and it is readily appreciated that the earthquake task of the Spring Valley Water Company was neither small from an engineering standpoint nor enviable from the municipal and political side. The destruction to the Spring Valley Water Company's plant as outlined above was produced by nothing less than a cataclysm, something which the mind of man could not foresee, and whose effects no engineering structure no matter how good could resist. The Crystal Springs conduit was of excellent design, but it was doomed on the marshes. The Pilarcitos pipe line had to succumb; it was right on the line of fault for a length of six miles. In the future it will be wise to avoid marshes and made ground for important pipe lines, and flexible joints should be introduced at intervals. The city was saved a terrible water famine simply because the San Andreas conduit, with slight repairs survived the general destruction. For safety the city needs a number of sources of water, in localities widely separated and not in the same geological region, with a number of main conduits so arranged that they will not tend to be destroyed all at the same time. The Spring Valley conduits have answered these requirements in so far as one of their number survived.

FIG. 38—*Diagonal Rupture, Pilarcitos Pipe Line, Spring Valley Water Company.*

Buildings and other structures in San Francisco to a great extent have been notoriously poor. The same is true in other cities in the earthquake belt. The time is ripe for the people to realize that they must enforce proper building laws and a proper attitude toward healthy construction both in municipal and in private works. The works of the Spring Valley Water Company are relatively of an exceptionally high type of construction. Its wrought iron conduits after thirty years of use, even in their present demolished condition, exhibit surprising preservation. Its pumping stations have survived where nearby structures collapsed. The Crystal Springs dam needs no praise. The earth dam at San Andreas fulfills its functions as well as ever, although it was directly on the line of the main fault and was greatly scarred. Faulty work and weak engineering construction may be found in other States of the Union, as well as in California. The engineer and contractor are not alone to blame, and I am not willing to criticize them. The community itself is partly responsible, because it expects and demands too much for its money.

FIG. 39—*Collapse by Compression at a Small Trestle, Pilarcitos Pipe Line, Spring Valley Water Company.*

I have already alluded to the destruction of water pipes within the city. Had the main conduits remained intact, there would still have been great difficulty in fighting the fire. Small reservoirs within the confines of the city should be connected with the main conduits by pipes of considerable size in no way connected with nor dependent upon the gridiron system of the streets. Had the city reservoirs of San Francisco tapped the large conduits independently of the street mains, some of the delay in obtaining water and fire pressure in the first fortnight after the earthquake might have been eliminated. Within the city confines also there should be larger reservoirs than are now provided.

Within the city boundaries, main branch pipes, and in general the pipes of the gridiron system were much destroyed. In the softer and made ground this was especially true. Moreover, the great extent of the fire destruction left innumerable tap and service pipes to hundreds of burned buildings in a most dilapidated condition. Explosions of gas mains, Fig. 40, added further rupture to the streets and the pipes beneath them. With this general demoralization of the gridiron system within the city and the loss of the Pilarcitos and Crystal Springs conduits, the situation on the morning of the earthquake may be understood. There was little water in the city and no water pressure.

Earthquakes are not uncommon in California and they will naturally occur again. There has been much talk of tapping the water sources of the Sierra Nevada Mountains and bringing that water to San Francisco by conduits and water courses which must be nearly 200 miles in length. As already stated, it is to be desired that the city have a number of distinct sources of water, but in the light of our present catastrophe how much more danger must there be of earthquake destruction upon a line of so extended a length? Conduits in duplicate would be of no avail; when one breaks so will the other. The only safeguard will be two distinct conduits running in widely separated districts, but such a proposition would entail great cost.

FIG. 40—*Gas Main Explosion, Valencia Street, near Market Street, San Francisco.*

FIG. 41—*Arcade, Inner Quadrangle, Stanford University.*

Palo Alto

Palo Alto is seven miles east of the fault line; its important earthquake damage occurred on the Campus of Stanford University. The extent of this destruction was great. The grade of workmanship on the University buildings was of relatively high quality, especially for the older buildings. The original buildings were planned with regard to earthquake possibilities. Steel-frame construction on the Pacific Coast was in its infancy at the time, and reinforced concrete almost unknown everywhere, therefore the earliest Stanford buildings and the inner arcades were built by day labor, with heavy cut stone and mortar rich in cement. The outer arcades and buildings, built later, were faced with ashlar and backed with rubble, too often improperly mixed with mortar. The structures built since 1902 were mostly unfinished in April, 1906, and were made of brick, thinly veneered with stone. A number of them contained some steel framing in domes or roof trusses, which, in vibrating, only helped to throw down the surrounding masonry walls. The earliest structures built remained essentially uninjured, except where they were too high, as in the case of the church and arch, but the latest additions were severely wrecked or destroyed, as they deserved. The central part of the Museum and Roble Hall were built of reinforced concrete, being among the first instances where that material was used. This construction was uninjured except for the falling of plaster.

I attribute the excessive destruction to three causes: (1), the nature of the ground; (2), the nearness to the fault line, and (3), the type of construction. I consider the third reason by far the most important. Heavy stone-faced buildings of the type we find at Stanford, no matter how well built, cannot resist an earthquake of the intensity experienced at Palo Alto, when we consider the nature of the ground upon which the buildings rest. Top-heavy stone arcades and walls, supported by light masonry pillars, see Fig. 41, are not constituted to withstand vibration, especially when the roofs are covered with heavy tile. The interior framing, especially of the later Stanford buildings, generally lacks the unity and stiffness which can come only through steel-cage construction or the best type of

FIG. 42—*Rear View Memorial Arch, Stanford University, Showing Arch Ring Intact.*

reinforced concrete cage construction. Considering earthquakes, I do not think the main type of structure at Stanford was happily chosen. The type lacks unity and coherence of frame. The buildings are rigid and often too high. The fault is more in the type of design, and not so much in the execution.

I have seen many pictures of the Memorial Arch showing dramatic destruction. These pictures give a false impression. Fig. 42, a view of the rear elevation, shows that the arch ring itself was not thrown down; in fact, there were but a few cracks in the crown of the soffit. It was the heavy top and cornice, insufficiently tied, that fell, as they should. It would have been surprising had they not fallen. Above the arch was a large box-shaped mass of masonry upon which rested the heavy overhanging cornice, and this box had practically no transverse framing or partition walls of any kind. The masses that dropped from the arch fell in large pieces, nearly 100 feet to the pavement without crumbling, Fig. 43, which proves that the mortar was good. Considerable cement was used in the mortar for the

Memorial Arch. Perhaps not so much, but at least a goodly amount of cement, was used in every building which I examined.

FIG. 43—*Fallen Masonry from Top of Memorial Arch, Stanford University, Showing Mortar Strength.*

In my opinion, had Stanford's buildings been of reinforced concrete, or of a good type of light brick or stone construction, bound to a yielding but unified metal frame, there would have been practically no destruction. The type of building should emphasize coherence and lightness, combined with yielding and elasticity. Moreover, it is a mistake to erect university and school buildings more than two stories in height; it is unwise for earthquake conditions and inappropriate for halls of assemblage.

FIG. 44—*View at Agnews Showing the Complete Collapse of the Main Tower of the Asylum.*

Agnews Asylum

Six miles to the north of San José one finds the Agnews Insane Asylum, consisting of one large brick, timber framed, building about two city blocks length, surrounded by some twenty similar but smaller brick structures. One hundred patients and eleven help were killed the morning of the earthquake. I do not know how many were injured. Not one of all these buildings remained habitable. Of all the destruction that I saw, and I visited the whole disturbed area, this cluster of buildings exhibited the most complete earthquake destruction, with the possible exception of the City Hall buildings in San Francisco. They are both public structures. Is it not time for California seriously to realize the situation? The main central brick structure and tower of Agnews Asylum, Fig. 44,

Fig. 45—*Earthquake Destruction, Native Sons' Hall, San José.*

entirely collapsed. The central tower fell en masse and crumbled to pieces so that three distinct wings of the buildings, themselves much demolished, were left disconnected. Some of the outer, smaller buildings had nine-inch brick walls for a height of two stories or more. It is significant to observe that a high water tower of structural steel, situated close to the power house of the asylum, remained entirely intact. The power house, built of brick, and its high brick chimney, collapsed with the rest of the brick structures.

San José

San José, about forty miles to the south and east of San Francisco, is thirteen miles to the east of the fault line. This city fortunately was spared a conflagration. The earthquake destruction was appalling, as is shown by a few typical photographs; see Figs. 45, 46, 47, and 48.

FIG. 46—*A View on First Street, San José, Showing Lack of Framing for First Floor. These Buildings Were Later Pushed Back into Place.*

San José water works, like that of Santa Rosa, was not injured; its sewers also were left intact, showing that there was no unequal displacement of the ground. The earthquake destruction was the result of severe vibration of poorly constructed brick and stone buildings. Again we find cheap construction with lime mortar, weak framing and insufficient anchoring for floors and roofs. Whole sides and fronts of two and three story store and office buildings on the main business streets collapsed; Fig. 45. On First Street, in the main business center, in one instance, a row of about six or seven buildings careened, due to a lack of sufficient transverse framing in the first story. Stores occupied these main floors, demanding large window openings and as much floor space as possible. Virtually the upper portions of the buildings rested on stilts; see Fig. 46. Public buildings in San José, with the exception of the postoffice,

FIG. 47—*Hall of Justice, San José, Completed in 1905.*

were generally racked. The tower of the postoffice was thrown down. In too many instances we find heavy stone facings improperly anchored to the interior frames, and roofs without system. In the great area of destruction, one finds roofs depending for their support upon the stiffness of the walls upon which they rested, while from the nature of design and construction the walls lacked stiffness. These flimsy or clumsy roofs, when subjected to earthquake vibration, tended to lower in the peak and to spread at the ends. Naturally they crushed out the weak outer walls, lacking in bond and inherently weak in mortar. Nearly all the public buildings, most of the school houses, and many churches in San José, where built of stone and brick, were demolished or severely damaged. Some of these structures still stand, but if wisdom is exercised, most of them will be torn down. Even some frame buildings, notably churches and the annex to the Hotel Vendome, improperly framed and with weak underpinning, completely collapsed; see Fig. 16.

Fig. 48—*Rear View, Hall of Records, San José, Showing How the Heavy Stone Outer Walls Cracked Away from the Interior Framing Due to Improper Anchoring of Walls to Floors and Roof.*

Pajaro Bridge

The Pájaro Bridge is perhaps the most interesting structural example of violent earthquake effect. It is on a six-degree curve whose chord makes an angle of about 40° with the fault line. The fault crosses the bridge near the west end. The bridge consists of two end deck-plate girder spans of fifty feet each, and four intermediate deck Pratt trusses, each of 120-feet span; Fig. 49. The abutments and the intermediate piers are of concrete with granite copings. This structure, in my judgment, would not have been affected by the earthquake had it been at a reasonable distance from the fault line, but it was subjected to a most severe racking, since the ground upon which its piers rested moved unequally. The distance between the end piers was increased 3½ feet. All the piers and abutments were moved more or less, but the heaviest movements occurred at the west end of the bridge. Fig. 50 shows the west abutment where the steel plate girders were moved 24 inches off the abutments and had to be supported by a temporary wooden bent. The plate girder span resting on this abutment did not fall during the earthquake, although it was dragged off the

FIG. 49—*Pájaro River Bridge, Southern Pacific Railroad; Temporarily Strengthened by Timber Bents and False Work; View Taken May 30, 1906.*

bridge seat, because it was help up by the rail fastenings and the riveted connections to the next span to the east. Fig. 51 shows the first intermediate pier at the west end of the bridge. This pier near its base was ruptured in a horizontal section, probably a level between two days' concrete work. The mass of the pier above the rupture was moved eastward about three inches with respect to the base. The coping was ripped from the top and moved eastward relatively more than 20 inches. The segmental rollers and the steel pedestals upon the coping were smashed and the anchor bolts severely bent. The next intermediate pier leaned somewhat to the east, and the anchor bolts above the coping were bent so that the pedestals of the steel work were moved some three inches. The distortion was less as one approached the eastern end of the structure. The eastern abutment was considerably cracked. The

FIG. 50—*West Abutment, Pájaro River Bridge, Southern Pacific Railroad.*

FIG. 51—*First Intermediate Pier, West End, Pájaro River Bridge, Southern Pacific Railroad.*

whole bridge superstructure was thrown slightly out of plumb. Earth motions are clearly seen in Fig. 50, where the sloping earth was moved 18 inches along the side wall of the west abutment.

Salinas Highway Bridge

The Salinas River runs through a marshy district of river deposits of slight elevation above the sea level. Although the region is at a distance of ten to twenty miles from the fault line, there was much evidence of differential surface movement, distinct from elastic vibration, just as in many other instances of soft surface deposits, for example, the San Bruno marsh near South San Francisco, where the Crystal Springs pipe line of the Spring Valley Water Company was so badly ruptured. The land on the south bank of the Salinas River, for a considerable area, was moved into the river in a northerly direction a distance of about six feet. Fig. 52 shows the south abutment of the Salinas highway bridge. The ground under the superstructure moved about six feet, careening the pile foundations as shown in the picture, without seriously injuring the trusses. A three-inch oil pipe line which crossed the bridge was ruptured on the south approach, one length of pipe being bent to form the letter "S." The northern approach to the bridge was hardly affected. In fact, the northern bank of the river shows little disturbance at this point. From the Salinas bridge eastward, a distance of about 1½ miles to Spreckels, the south bank of the river is continuously scarred and rent so that a wagon road was made impassable for vehicles.

The Spreckels Sugar Mill

At Spreckels is found the Spreckels Sugar Mill. The mill buildings are of excellent construction. The main building, about 100 × 500 feet, a huge box-shaped structure with heavy steel frame, brick walls supported on the frame, and arched concrete floors with steel floor framing, rests on a pile foundation. The mortar for the brick work is excellent and contains considerable cement. Had these buildings been in Santa Rosa or San José, I think they would have stood the

FIG. 52—*South Abutment, Salinas Highway Bridge.*

earthquake with possibly a few slight cracks and the loss of some brick here and there near the cornices and corners. But unfortunately they rest on soft ground, very much like the made ground areas of San Francisco. Considerable portions of the brick walls were thrown out of the steel frames. The long walls of the main building are supported by steel columns about every 15 feet and are braced transversely at these columns by the interior framing; except in the middle third of the building, where there is an open space from ground to roof to give clearance to the huge machinery. In this middle third, where the long walls are not laterally braced by interior transverse frames and floors, the walls were badly buckled.

The east wall was buckled concave outward, and was badly cracked, but little of it fell. The west wall, in the middle third, was buckled convex outward, and naturally the brick work was thrown down in large masses. Fig. 53 shows the west wall of the building. I believe this building was subjected to an earth movement from south to north, which accounts in a measure for the buckling of the long

FIG. 53—*West Wall, the Spreckels Sugar Mill, near Salinas, Showing Bulged Wall and Brickwork Thrown from the Steel Frame.*

FIG. 54—*North Wall, the Spreckels Sugar Mill, near Salinas, Showing Ruptured Brick Curtain Walls and the Advantages of a Class A Steel Frame.*

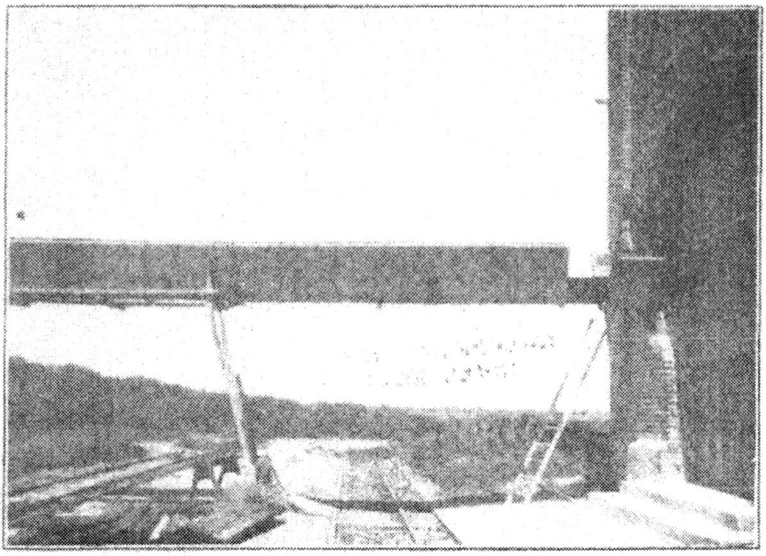

Fig. 55—Ruptured Conveyor, the Spreckels Sugar Mill, near Salinas, Showing Slipping of the Ground into the River.

walls in the middle unsupported third. There may have been some settling of the walls also. Large masses of brick in the north wall, Fig. 54, were thrown down. Fig. 55 shows a rupture in the carriage way to the dumps at the south end of the structure, clearly showing the existence of a decided ground movement at the time of the earthquake. All the smaller buildings were damaged to some extent. The Spreckels Sugar Mill was well designed. It was probably subjected to a more severe racking than any other well designed structure in the State. It can readily be repaired and speaks for the stability of Class A construction.

Conclusions

There has been much comment of late regarding the relative merits of sea level and lock types of canal for the Isthmus of Panama, in the light of earthquake possibilities. From a study of the earthquake destruction to San Francisco buildings it would appear that honestly, scientifically and generously constructed works can abundantly withstand any reasonably severe earthquake shock so long as the structure does not happen to exist on a fault line or a place of very unequal motions of the ground. On a fault line no structure can fully withstand the shock, no matter how well built. If fortunately located, however, a structure on or near a fault line may not be seriously crippled. In their severe locations it is seen that the 93-foot San Andreas earth dam still fulfills its functions and the Pájaro bridge, though badly crippled, was speedily made usable; the Spreckels Sugar Mill was repaired without great expense, and large well built buildings on honestly designed and deep foundations stood the racking on the made ground in San Francisco. In the future, water supply conduits should avoid fault lines, marshes, and other unstable ground whenever possible. Dams should not cross fault lines, and when possible should not be run at right angles thereto; yet considering their purpose it usually will be necessary to place dams in the most unsatisfactory positions. It is not wise in earthquake countries to build bold arch dams of thin construction. Important railway bridges should be located at river crossings not coincident with lines of crustal weakness, and should avoid soft alluvial deposits. It is easy to state this advice and these restrictions, but it is another thing to follow them. The locating engineer is not always free to choose his site. Where structures must be built upon treacherous ground or near fault lines, no expense should be spared for good materials, high grade workmanship, and intelligent design. The most generous factors of safety should be used and the best designing talent employed. California engineers can not pay too much attention to geologic structure.

Careful examination of all types of structures within the earthquake belt leads me to emphasize the following detailed conclusions. I do not expect every one to agree with them. Some of the statements may be very radical, and some may be wrong:

1. Cornices and top walls of buildings, especially of brick and stone structures, should not be heavy nor have great projection. In the future cornices and top walls should be more securely anchored with metal, their masonry bond should be made with care and the

cementing materials should be of high quality. A considerable number of lives were lost by falling brick and stone from such sources. These parts of buildings have been very generally demolished by the earthquake.

2. Cornices and top walls of first class steel cage constructed buildings, where the steel framing and anchoring has been carried into the cornice work, have resisted the earthquake.

3. Projecting brick work and fire walls of ordinary brick and stone structures have proven themselves an abomination, especially where lime mortar was used. With cement mortar the destruction was nearly as great, the only difference being that the material fell in large masses.

4. Brick chimneys for ordinary brick or frame dwellings should be built of weak lime mortar above the roof line. This seems a paradoxical statement. In a severe earthquake the brick chimney will not move with the house. It must collapse.

If built with rich cement mortar it will fall in one piece and crush through the roof. If built of lime mortar it will crumble and the individual bricks will roll off the sloping roof. Metal chimneys should be studied for the part above the roof line. If brick is insisted upon from preference or custom, reinforce the brickwork with steel rods and bands.

5. Terra cotta and similar materials should not be used so profusely and boldly in the future. They are not good earthquake materials. They have been badly cracked by earthquake vibration. The north wall of the Fairmont Hotel, where it is faced with terra cotta, was badly cracked. Pressed brick work with good mortar and good bond is a far more satisfactory material. Terra cotta ornamentation on the Mills Building and many other buildings was badly spalled by vibration. I would not eliminate terra cotta from use, but it should

be less highly ornamented, be manufactured with greater thickness, and the hollows should be filled with a cementing matrix.

6. Hollow tile partitions in the General Postoffice were almost wholly destroyed by the earthquake. Of course this building rested upon very treacherous ground, but the earthquake certainly did very much more damage to tile partitions in San Francisco than is generally admitted. These partitions have little strength and are readily collapsible. They have no elastic continuity.

7. I believe hollow tile floors were less disturbed by the earthquake than tile partitions, because there was less tendency to unequal motion in the plane of the floor; whereas partitions were subject to a greater racking motion. Tile partitions and floors do not appeal to me with the same force as reinforced concrete construction, especially when the question of fire-proofing is left out of the discussion.

8. Brick buildings without steel work and of light construction should have small height, three to four stories at most. Their bond of brick work should be carefully inspected and their brick walls should be securely tied to the floor and roof frames. On soft ground their foundations should be even more carefully designed for stiffness of framing and distributing power than those for wooden buildings. The bond of brick work in San Francisco was proverbially bad.

9. Cheap lime mortar should not be allowed for buildings in the congested districts of San Francisco. Cement with just enough lime to make it workable—say one part cement to four or five of lime mortar—should be more generally insisted upon, not only in the congested parts of the cities, but everywhere where brick buildings are used.

The high school building at Berkeley may be taken as an example of lime mortar building. It was most seriously cracked, especially in the second story. Hundreds of buildings might be named in this connection. In the light of what has happened it is a crime to use bad bond and lime mortar for brick work in schools and public buildings, in fact in all buildings. Mortar has been too generally applied to dry brick.

10. Brick buildings of greater height than four stories should have heavy walls and a sufficient number of interior cross walls to give lateral stiffness. The Appraiser's Building or Custom House was unharmed by the earthquake. The Palace Hotel stood the shock splendidly, the latter being an excellent type of brick structure, whose brick walls were reinforced with iron rods. The foundations of such buildings should be designed with the greatest care to prevent unequal settlement.

11. All over the earthquake belt one found brick and stone structures badly demolished by the falling out of main walls due to improper design of the roof trusses. These trusses too generally were of clumsy wood construction improperly anchored to the supporting walls and unscientifically framed. Too generally they were lacking in a lower chord tension member. The earthquake caused such roofs to spread and kick out the walls, which had no ability whatever to resist a lateral thrust. The high school building in Berkeley is a typical example. For large school buildings roof trusses of steel should be used. One saw many cases where brick walls stood the shock that would have been thrown down except for the tying property of the roof trusses above them. The Majestic Theatre in San Francisco clearly showed how steel trusses kept high walls from collapsing.

12. Properly constructed wooden buildings withstood the earthquake with the exception of their brick chimneys, no matter where located in the earthquake belt, except on the fault line and in regions of the greatest disturbance on soft ground. Some frame buildings collapsed upon favorable ground due to improper underpinning. There should be more continuity in the frames of

wooden buildings, especially at the floor levels, and the underpinning should be more carefully attended to in the future.

13. At the time of the earthquake there were entirely too many top-heavy and improperly braced brick and stone towers and steeples in San Francisco and other cities. Where they are merely ornamental, mere masonry, towers should be discouraged as far as possible. Their ruins were everywhere to be seen in San Francisco, Oakland, Santa Rosa and San José. The Ferry tower might have had heavier framing, although it should hardly be spoken of in this connection. Most towers lack in interior cross walls and in necessary steel frames. The Memorial Arch at Stanford University has been referred to. Reinforced concrete, where not too boldly employed, is well adapted for these purposes.

14. Important buildings like the Postoffice should not be placed on filled ground or treacherous ground, but when they must be so placed, they should have heavier steel framing and lighter masonry work than the Postoffice. The type of building structure to be selected in these cases should be determined by the nature of its foundation site. Remarks which I have made concerning buildings at Stanford University are here pertinent.

15. High buildings with deep pile foundations of the proper design withstood the earthquake shock well on soft ground. Considering the nature of the material on which the Ferry Building rests, it stood the shock splendidly because of its excellent pile foundations. The Call Building foundation represents the slab type, apparently equally well fitted for service.

16. Reinforced concrete cage construction should be more respected in the future by the building laws and trades unions of San Francisco. There is no reason why buildings of this type, designed by competent engineers, should not be at least six or eight stories in height.

17. Low buildings of intelligent reinforced concrete construction are far more able to resist earthquake shock than brick and stone buildings.

18. On soft ground the footings of ordinary buildings too light to require pile foundations might have cellar slabs of reinforced concrete to act as units with the wall footings to give distributing power and prevent unequal settlement.

19. Reinforced concrete sewers should be studied in the light of brick sewer destruction in the made ground of San Francisco.

20. Important water mains should avoid soft ground and when they must necessarily pass from firmer to softer ground they should be provided with flexible joints and cut-offs.

21. Important water mains on which the fire service depends, which must run through the made ground of the city, should be of riveted wrought iron or steel, have flexible joints at intervals and be lodged in tunnels, say of reinforced concrete. Earthquake disturbance near the water front might severely crack such tunnels without great injury to the pipes within, due to the properties of the materials, the nature of the pipe joints, and the clear space between the tunnel walls and the pipe. Such a construction would further give more probability of access to the pipe in case of calamity. Important networks of pipes in the gridiron system should be arranged in more or less independent units with respect to the softer and firmer grounds of the city; so that flow in the pipes on made ground could be quickly separated from that on more solid foundations.

22. Main conduits running from storage reservoirs to the city should avoid marshy lands as far as possible. Where they must necessarily cross swamps and marshes, they should be provided with flexible joints and not be too firmly blocked to their platforms.

23. The city should have a number of sources of supply in widely separated localities of distinct geological formation.

24. Equalizing reservoirs within the confines of the city should be numerous and of considerable capacity.

25. Small distributing reservoirs within the city limits should be connected with the main conduits by large pipes independent of the gridiron system. These pipes should be carefully designed and be easy of access. With such provision, some of the delay incident to the forcing of water through crippled street mains to the various city reservoirs and pumping stations might have been avoided. Artesian wells should be encouraged as local sources of supply.

26. The business district should be safeguarded by a salt water system in addition to the regular water supply for fire service, and where salt water pipes must run through made ground they should be provided with flexible joints and with a tunnel construction as has been suggested for the main water pipes; or they should be on the surface.

27. San Francisco needs a fire boat service.

28. High brick chimneys for factories should be built with cement mortar; even then they should be reinforced. In many cases a lower chimney with a forced draft should be considered. A study of reinforced chimneys should be made. They would withstand earthquake shocks much better than brick ones.

29. In the business section of the city Class A buildings and first class reinforced concrete structures should be encouraged at the expense of Class B structures, and the skimping of steel frames should meet with entire disapproval. Diagonal framing should be introduced wherever windows and interior passage ways or

openings do not prevent; better still, heavy knee braced framing and spandrel girder framing should receive more attention.

30. The design of high buildings with self-supporting walls should be discouraged.

31. Steel columns should run through more than one floor, and their splice joints should be strongly designed.

32. Cast iron columns should not be used.

33. Heavy stone ornamentation should not be hung to the steel frames of high buildings, and heavy centralized supports on the first floor, as in the Flood Building, should be avoided when possible. No expense should be spared for foundations of buildings of great height.

34. Heavy stone corner piers at the sidewalk level of steel framed buildings should rest upon the frame and not upon masonry walls under the sidewalk. Where such piers rested upon concrete under the sidewalk, the stone work was badly cracked; where they were supported by the steel frame no cracks were found.

35. Face brick should be carefully bonded to the back brick in Class A buildings. Large areas of face brick fell from the west wall of the Merchants Exchange Building because of improper bond.

36. Reinforced concrete buildings of careful, honest, and intelligent design should be allowed to enter competition in San Francisco. It is unfortunate that reinforced buildings have been placed in the Class B list. To the layman this implies inferiority to Class A structures. Of their types Class A buildings are not to be considered better than first-class reinforced concrete cage constructed

buildings, only I should not use the latter type for structures of the greatest height.

37. Neglecting the problem of surface appearance, Class A buildings might be designed with curtain walls of reinforced concrete instead of using brick and terra cotta.

38. High buildings like the Mills Building in which the floors are supported on a steel frame, but whose walls are self-supporting, should not be imitated in design in the future. When such buildings are badly shaken, or when their outer walls are badly damaged, repair is difficult. In Class A buildings, where the steel frame carries the load of each floor independently, such difficulty vanishes. The Mills Building had the further miserable feature of columns breaking joints at every floor.

39. Electric insulation for high tension transmission should be rigorously inspected. Chimneys on cheap brick and frame dwellings should be more sensibly built. Electric wires and bad chimneys were fruitful sources for the starting of fires immediately after the earthquake.

The Investigation of the California Earthquake

By Grove Karl Gilbert

Of the U. S. Geological Survey

Published by permission of the director of the United States Geological Survey and of the chairman of the California Earthquake Investigation Commission.

It is the natural and legitimate ambition of a properly constituted geologist to see a glacier, witness an eruption and feel an earthquake. The glacier is always ready, awaiting his visit; the eruption has a course to run, and alacrity only is needed to catch its more important phases; but the earthquake, unheralded and brief, may elude him through his entire lifetime. It had been my fortune to experience only a single weak tremor, and I had, moreover, been tantalized by narrowly missing the great Inyo earthquake of 1872 and the Alaska earthquake of 1899. When, therefore, I was awakened in Berkeley on the eighteenth of April last by a tumult of motions and noises, it was with unalloyed pleasure that I became aware that a vigorous earthquake was in progress. The creaking of the building, which has a heavy frame of redwood, and the rattling of various articles of furniture so occupied my attention that I did not fully differentiate the noises peculiar to the earthquake itself. The motions I was able to analyze more successfully, perceiving that, while they had many directions, the dominant factor was a swaying in the north-south direction, which caused me to roll slightly as I lay with my head toward the east. Afterward I found a suspended electric lamp swinging in the north-south direction, and observed that water had been splashed southward from a pitcher. These notes of direction were of little value, however, except as showing control by the structure of the building, for in another part of the same building the east-west motion was dominant.

In my immediate vicinity the destructive effects were trivial, and I did not learn until two hours later that a great disaster had been wrought on the opposite side of the bay and that San Francisco was in flames. This information at once incited a tour of observation, and thus began, so far as I was personally concerned, the investigation of the earthquake. A similar beginning was doubtless made by every other geologist in the State, and the initial work of observation and record was individual and without concert. But organization soon followed, and by the end of the second day it is probable that twenty men were working in co-operation under the leadership of Professor J. C. Branner, of Stanford University, and Professor A. C. Lawson, of the State University at Berkeley. At that time and for several succeeding days the ordinary means of communication were so paralyzed or overburdened that no messages passed between these two centers of organization; but as the needs of the hour were patent to all, the work was not prejudiced by the lack of intercommunication.

On the third day after the shock Governor Pardee appointed a State Earthquake Investigation Commission, naming as its chairman the head of the geological department of the State University, Professor Lawson, and including in its membership Professor Branner, of the Stanford University, Professors Davidson and Leuschner, of the State University, Professor Campbell, of the Lick Observatory, Mr. Burckhalter, of the Chabot Observatory, Professor Reid, of Johns Hopkins University, and Mr. Gilbert, of the United States Geological Survey. The commission held its first meeting three days later, when the scope of its work was considered and defined, provision was made for circulars soliciting information, an announcement was prepared for the purpose of relieving certain groundless fears entertained by a portion of the community, and two committees were appointed for the general work of observation. To the first committee, with Professor Lawson as chairman, was assigned the determination and study of surface changes associated with the earthquake and the collection of data as to intensity, so that isoseismals, or curves of equal intensity, might be drawn upon the map. To the second committee, with Professor Leuschner as chairman, was assigned the collection of data for the drawing of coseismals, or lines connecting points on the earth's surface reached by the shock at the same instant. Some weeks afterward, when the main features of the earthquake had become known, a third committee was appointed, with Professor Reid as

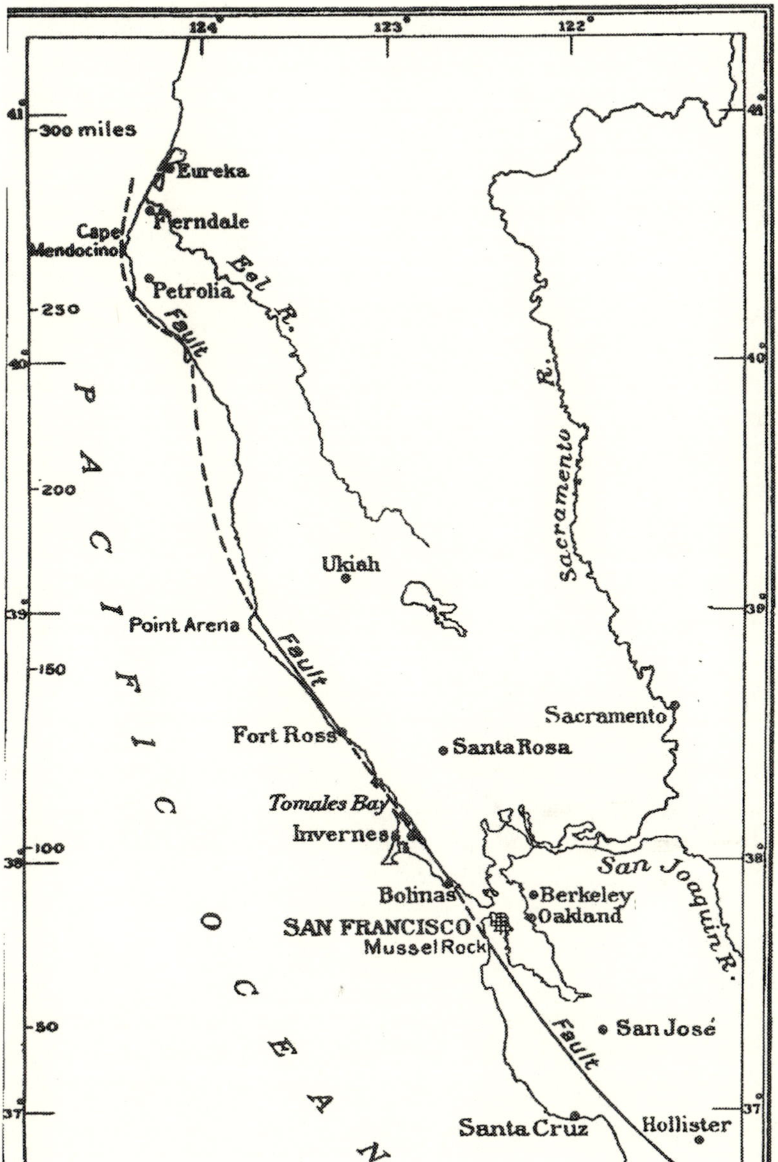

FIG. 1—*Map Showing the Position of the Fault Which Caused the California Earthquake.*

chairman, to consider the relations of the earthquake phenomena to certain problems in geophysics, or the science of the inner earth. The work of these three committees is still in progress, and will not be completed for several months. The actual drawing of isoseismals and coseismals can not be performed until a large body of observations have been compiled and studied, and the geophysical problems are as yet only imperfectly formulated; but of the physiographic phenomena, or the disturbances of the earth's surface, so much is known that it has been thought advisable to prepare a preliminary report. This was submitted to the governor on the third of June, and has been issued as a pamphlet of twenty pages. The expenses of the commission are being met by the Carnegie Institution.

Architects and engineers were not less prompt and energetic. To the men who plan and direct construction in the earthquake district of California it was important to know what materials and what structural forms best withstood the shock, and they immediately began the study of earthquake injuries and of instances of immunity from earthquake effects. In that part of San Francisco where the earthquake injury was most serious the shock was quickly followed by fire, which destroyed much of the evidence, but many important observations were made in the brief interval. The study of structural questions, like the study of natural phenomena, was at first individual only, but afterward was aided by organization. Committees were appointed by various professional societies, national and local, and were charged with the investigation of specific structural questions, and the results of their labors will find place not only in the transactions of the societies, but in revised building regulations and in important modifications of municipal plants for lighting and water supply. Various bureaus of the national government have also taken part in the structural studies, sending experts to San Francisco and other localities of exceptional earthquake violence.

FIG. 2—*Fault Topography between Tomales and Bolinas Bays: Looking Northwest. The General Slope Toward the Left Has Been Interrupted by a Slight Uplift of the Part at the Left. The Pond Occupies a Hollow Thus Produced.*

The Japanese government promptly sent to California a committee of investigation headed by Dr. Omori, professor of seismology in the University of Tokyo, and composed otherwise of architects and engineers. The first conference of these visitors with the State Commission warranted the suggestion that we may find it as profitable to follow Japanese initiative in the matter of earthquake-resisting construction as in that of army hygiene.

The following sketch of the physical features of the earthquake is based chiefly on the body of data gathered by the State Commission:

An earthquake is a jar occasioned by some violent rupture. Sometimes the rupture results from an explosion, but more commonly from the sudden breaking of rock under strain. The strain may be caused by the rising of lava in a volcano or by the forces that make mountain ranges and continents. The California earthquake of April 18 had its origin in a rupture associated with mountain-making forces. A rupture of this sort may be a mere pulling apart of the rocks so as to make a crack, but examples of that simple type are comparatively rare. The great majority of instances include not only the making of a crack but the relative movement or sliding of the rock masses on the two sides of the crack; that is to say, instead of a mere fracture there is a geologic fault. After a fault has been made its walls slowly become cemented or welded together, but for a long time it remains a plane of weakness, so that subsequent strains are apt to be relieved by renewed slipping on the same plane of rupture, and hundreds of earthquakes may thus originate in the same place. From the point of view of the geologist the displacements of rock masses are the primary and important phenomena; the faults are incidental phenomena, of great value as indices of the displacements; and the earthquakes are of the nature of symptoms, serving to direct attention to the fact that the great earth forces have not ceased to act.

A faulting may occur far beneath the surface and be known only by the resulting earthquake; but some of the quake-causing ruptures extend to the surface and thus become visible. The New Madrid and Charleston earthquakes are examples of those having deep-seated origins, the Inyo and California, of those whose causative faults reached the surface of the ground.

The general character of California earthquakes was so well known that when the dispatches told of a severe shock at San Francisco no American geologist had a moment's doubt that it was caused by a fault movement, and among those specially conversant with the

Fig. 3—*This Fence, Previously Continuous and Straight, Was Broken and Parted by the Earthquake Fault, the Offset Being 8½ Feet. The Line of Fault, Concealed by the Grass, Crosses the Ground from Left to Right, Touching Both the Dissevered Ends of the Fence.*

structure of the affected district attention was immediately directed to several fault lines, with the expectation that one or more of them would show the marks of fresh dislocation. Mr. Ransome prepared a prophetic article in which he indicated the lines most likely to be concerned. *Nat. Geog. Mag., Vol. 17*, 1906, pp. 280-296. Professor Branner stated in an interview that he had immediately made a forecast of the locality of the origin and that it had proved to be correct, and Mr. Fairbanks went at once to a zone of "earthquake topography" with which he was already acquainted, and found a fresh rupture in the expected place.

The California earthquake was caused by a new slipping on the plane of an old fault which had been recognized for a long distance in California, and in one place had been named the San Andreas fault. Associated with this fault is a belt of peculiar topography, differing from the ordinary topographic expression of the country in that many of its features are directly due to dislocation, instead of being the product of erosion by rains and streams. One of its characteristics is the frequent occurrence of long lines of very straight cliffs. Another is the frequent occurrence of ponds or lakes in straight rows. The tendency of erosion is to break up such cliffs into series of spurs and valleys and to obliterate the lakes by cutting down their outlets or filling their basins with sediment. Fig. 2 shows

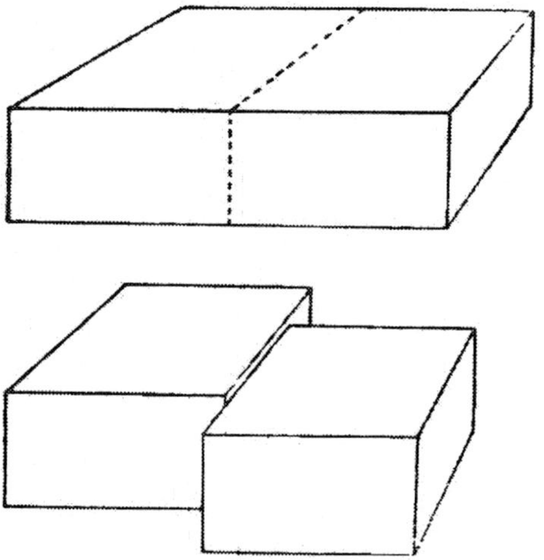

FIG. 4—*Diagrams Illustrating the Dislocation Causing the California Earthquake. The Upper Represents an Earth Block 100 Feet Square and 25 Feet Thick, with Indication of the Position of the Fracture. The Lower Shows the Relation of Its Two Parts after Faulting.*

one of the fault-made ponds. This line and zone have been recognized by California geologists through a distance of several hundred miles. It was to this line that attention and expectation were especially directed, and it was on this line that the surface evidence of new faulting was actually found. The new movement was not coextensive with the line as previously traced, but affected only the northwestern portion; and, on the other hand, it extended farther to the northwest and north than the old line had previously been recognized. The map represents only the line along which the recent change occurred. From a point a few miles southwest of Hollister it runs northwestward in a series of valleys between low mountain ridges to the Mussel Rock, ten miles south of the Golden Gate. Thence northwestward and northward it follows the general coast line, alternately traversing land and water. The farthest point as yet definitely located is at Point Delgada, but the intensity of the shock at the towns of Petrolia and Ferndale probably indicates the close proximity of the fault and warrants the statement that its full length is not less than three hundred miles. South of Point Arena its course is direct, with only gentle flexure, but the data farther north seen to imply either branching or strong inflexion. Opposite San Francisco its position is several miles west of the coast line, and it nowhere touches a large town.

That which occurred along this line was a differential movement and permanent displacement of the rock and earth on the two sides of a vertical crack. The principal displacement was not vertical, but horizontal. If one thinks of the land to the east of the crack as stationary, then the change may be described as a northward movement of the land west of the crack. If the land to the west be thought of as stationary then the land to the east-ward moved toward the south. It is probable that both tracts shared in the movement, the eastern shifting toward the south and the western toward the north. Perhaps the nature of the change can be more readily understood by reference to Fig. 4, which represents an ideal block of the earth's crust, 100 feet square on the surface and 25 feet deep, before and after its division and dislocation by the earthquake-causing fault.

Fig. 5—A Faulted Road near the Head of Tomales Bay. The Nearer and More Distant Parts of the Road Were Originally in One Line—a Continuous, Straight Road. The Present Offset is Twenty Feet.

Wherever a fence, road, row of trees, or other artificial feature following a straight line was intersected by the fault its separated parts were offset, and an opportunity thus afforded for measuring the amount of change. The measurements range in the main from 6 to 15 feet and have an average of about 10 feet. At one place (Fig. 5) a road was offset 20 feet, but in this case the underlying ground was wet alluvium and part of its movement may have been due to a flowing of the soft material. There was also some vertical change, but this was not everywhere in the same direction and its amount was comparatively small. At many points the land west of the fault appears to have risen one or two feet as compared with the land at the east.

Fig. 6—*Ordinary Appearance of the Earthquake Rift Where It Traverses Firm Turf.*

The surface manifestation is not usually a simple crack, but a disturbed zone a few feet broad, the earth within the zone being split into blocks which show more or less twisting or rotation. In some places the zone is slightly depressed below the adjoining surfaces, and elsewhere slightly elevated. Other disturbances of the surface were associated with the earthquake, but the track of the central fault has a character of its own, a character with which the field workers soon became familiar, so that it could be clearly identified. It came to be distinguished in their conversation and note-books as "the rift." For considerable distances the rift is single, but elsewhere it is more or less divided, the parts lying within a few rods of one another and being approximately parallel. There are also branches parting from the main rift at various angles and gradually dying out in the adjacent country, and in some of these the belt of disturbance is broad and complicated (Fig. 7). There are also outlying cracks occurring within a mile or two of the central rift and having irregular courses, and these may probably be referred to the same general system of rock strains.

Fig. 7—A Zone of Earthquake Fracture Where It Crosses a Road Near Bolinas.

Other cracks are distinctly secondary in character; that is to say, they are not due directly to the stresses and strains by which the fault was made, but are results of the earthquake itself. The jar constituting the earthquake, or in technical language the earthquake wave, as it travels through rock and earth produces temporary compressions and other strains, and these often occasion cracks at the surface. Where the material is elastic such secondary cracks merely open and close, leaving the ground with its original form;

but where it is inelastic and incoherent, as in the case of young alluvial formations and artificial fillings, some of the cracks opened by passing waves do not close again, but remain as permanent vestiges of the shock. Closely associated with these secondary cracks in soft ground are permanent changes in surface form. At the head of Tomales Bay, for example, a broad tract of soft ground between high and low tide was thrown into low ridges, with cracks along their crests, and these remained until destroyed by wind waves. In San Francisco considerable tracts of "filled" land were shaken together and thus made to settle a few feet, and were at the same time slidden several feet toward the bay (Fig. 9).

FIG. 8—*Cracks Caused by the Shaking of Marshy Ground. The Comparatively Firm Road Embankment Preserved the Cracks Better than the Bog.*

Certain changes, very conspicuous to the observer who drove about the country, are closely associated with roads. A side-hill road is usually constructed by excavating a notch in the natural slope and piling the excavated material in an embankment at the outer edge of the notch. In course of time, and especially during rainy seasons, the embankment at the outer edge of such a road settles and has to be built up as a matter of repair. Portions of the bluff on the up-hill side of the notch are also apt to fall away, taking the form of small landslides, which have to be removed from the road as a rule after every rainy season. The earthquake precipitated many changes of this sort. Along all side-hill roads in the immediate vicinity of the rift a crack was developed between the embankment and the original soil against which it rested, and this crack often assumed formidable dimensions (Fig. 10); in fact its magnitude was found to be a convenient index of the local violence of the earthquake in regions where buildings are rare. Landslips from the bluffs margining the roads (Fig. 11) were also very numerous, in many instances stopping traffic until repairs could be made. And there were many landslides on a larger scale, the earthquake initiating movements which might otherwise have been delayed for years or even centuries. Some of these landslides fell into streams, dammed their waters and created temporary lakes.

FIG. 9—*Street Scene in San Francisco, Showing Effect of the Earthquake on Filled Ground.—The Distant Part of the Street Probably Retains Its Original Level and Position. Nearer by the Ground Has Settled Several Feet and Has Also Moved to the Left.*

Other disturbances of water supply were more directly connected with the earthquake. At several points large volumes of water were squeezed from the ground during the agitation, causing brief but violent torrents, and one of these brought with it so much sand as to constitute a sort of sand eruption. There are reports also that certain springs have received a permanent increase in volume, a result which would naturally follow from the modification of underground circulation by the cracking of rock and earth.

Fig. 10—*Road Crack Caused by the Earthquake.*

Wherever the shock was specially strong there was considerable injuries to trees; some were overturned, others broken near the ground, and yet others broken near their tops. A number of large redwood trees standing on the line of the rift were split from the ground upward, the basal portions being faulted along with the ground they stood on.

FIG. 11—*Small Landslide on the Uphill Side of a Side-Hill Road.*

In the systematic survey of the earthquake area the relative intensity is being estimated by means of the records of various physical effects. In the immediate vicinity of the fault road-cracks and cracks in alluvium are large and numerous; many trees were broken or overturned; there were many landslides; half of the wooden buildings of any village or hamlet were shifted horizontally, often with serious injury; buildings and chimneys of brick or stone were thrown down; during the shock men, cows, and horses found it impossible to stand, and fell to the ground; and some persons were even thrown from their beds. In a general way all these evidences of violence diminish gradually with distance from the fault on either side. The rate of diminution, with exceptions to be mentioned presently, may be expressed by saying that at five miles from the fault only a few men and animals were shaken from their feet, only a few wooden houses were moved from their foundations, about half the brick chimneys remained sound and in condition for use, sound trees were not broken, and no cracks were opened which did not immediately close. At a distance of twenty miles only an occasional chimney was overturned, the walls of some brick buildings were cracked, and wooden buildings escaped without injury; the ground was not cracked, landslides were rare, and not all sleepers were wakened. At seventy-five miles the shock was observed by nearly all persons awake at the time, but there were no destructive effects; and at two hundred miles it was perceived by only a few persons.

A number of exceptions to this gradation of intensity are connected with tracts of deep alluvial soil, especially if saturated with water, and with tracts of "made ground." The great destruction in the low-lying part of San Francisco, eight miles from the fault, is directly connected with the fact that much of the ground there is artificial, the area having been reclaimed from the bay by filling in with sand and other materials. The severity of the disaster at San José, twelve miles from the fault, has been ascribed to the deep alluvial soil on which the town stands, and many other local peculiarities seem to admit of the same explanation. It is necessary also to distinguish carefully between earthquake intensity and destructive effect, because injury to property was conditioned by mode and material of construction no less than by intensity of vibration. But after making due allowance for differences in natural foundation and for differences in the resisting power of buildings, there remain various anomalies for which satisfactory explanation has not as yet been found. The natural foundation of Oakland is similar to that of San José, and its distance from the earthquake origin is about the same, but the injury to its buildings was decidedly less; and Santa Rosa, standing on ground apparently firmer than that at Oakland or San José and having a somewhat greater distance from the fault, was nevertheless shaken with extreme violence.

It is too early to discuss these anomalies. With the data now in hand it seems to be true that there are outlying tracts of high intensity surrounded by areas of relatively low intensity; and these features, if they shall be fully established, will doubtless affect in some important way the general theory of the earthquake.

One of the chief uses of time observations in connection with most earthquakes has been to determine the position of the origin. As the elastic wave travels outward in all directions from the initial point it reaches successively points on the earth's surface which are more and more remote. Coseismal lines, or lines of simultaneous arrival, are, therefore, closed curves circling about the region of the initial fracture. In the case of the California earthquake this particular function of the coseismals is not required, because the fracture is visible at the surface; but they are not therefore without value. It is not to be supposed that the yielding of the earth occurred at the same instant throughout the entire extent of the fault plane. We should assume, rather, that the fracture, beginning at some point, was extended thence to the remainder of the tract, a certain amount of time being consumed in its propagation. When the time data have been collected and studied, it may be possible to determine at what point the fracture began and at what rate it was extended. It is hoped also that when the time records and intensity records shall have been systematically discussed there may result some conclusion as to the depth to which the fault extended and also as to its subterranean form.

FIG. 12—*Water Tank Thrown from Its Pedestal by the Earthquake.*

Mention has already been made of the question whether the permanent dislocation or change of absolute position involved in the faulting was divided between the tracts of land on the two sides or was confined to one or the other of them. At first sight it would appear that the only thing susceptible of actual determination is the relative displacement, and that the absolute displacement, or the real movement with reference to the earth as a whole, must remain a matter of theory only. Nevertheless, it happens that in this particular instance the real changes in geographic position are not only susceptible of determination, but are actually to be investigated. To illustrate the problem, let XY represent, in ground plan, a portion of the fault line, and let ABB'C be the original position of a straight line intersected by the fault. Assuming for the moment that the dislocation was equal on the two sides of the fault, then the line AB was carried to the position DE, and the line B'C to the position FG. We may think of the distances BE and B'F as each equal to five feet. The dislocation of five feet pertains to every point near the fault line, but it is not supposable that the same dislocation affects points at a great distance from the fault. At some remote

FIG. 13—*School-House at Point Reyes Station, Near Inverness; Shifted Horizontally Two and One-Half Feet by the Earthquake. The Corner Here Shown Was Slipped from the Foundation and Rests Directly on the Ground.*

point, for example Z, in the direction B'C, there was no displacement. If B'C and FG were both produced in that direction they would be found not precisely parallel, but would eventually coalesce. How far the undisturbed region Z may be from the fault line is a matter of pure conjecture, but we may plausibly assume that the transverse dimension of the area affected by the displacement is of the same order or magnitude as the length of the fault line and is measured by hundreds of miles. If this assumption is correct, then throughout a great region in central and northern California all points have experienced a change in geographic position, the change in the vicinity of the fault being of about five feet and the amount diminishing toward the northeast and southwest. If the only determinations of latitude and longitude within this area were of the ordinary approximate character, it would be impossible to measure the changes in geographic position theoretically accomplished by the fault; but it fortunately happens that the region is traversed by two belts of the triangulation of the United States Coast and Geodetic Survey, one being a system of triangles for the control of the coastal map work, and the other the elaborately measured transcontinental belt. The region thus contains several scores of points whose co-ordinates have been determined with a high degree of precision, and it is possible by the redetermination of these positions to measure the dislocations which have taken place in connection with the earthquake. As all topographic and hydrographic maps of California are dependent for their latitudes and longitudes upon the positions given by this triangulation, and as there is reason to believe that many of these positions have been disturbed by a measurable amount, the superintendent of the Coast Survey has determined to repeat so much of the work of triangulation as may be necessary in order to redetermine the geographic positions. And it is proposed to carry this work far enough eastward to connect the redetermined points with stations that may safely be regarded as quite beyond the effect of the recent fault. When this has been accomplished much light will be thrown on the nature and distribution of the strains which were relieved by the dislocation along the fault line, and it will be possible to say definitely whether the original displacement involved the territory on both sides of the fault or on one side only.

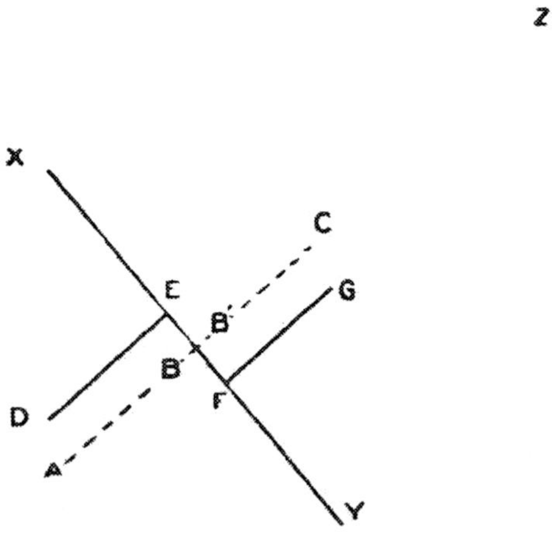

Fig. 14—*Diagrammatic Plan of a Portion of the Earthquake Fault, Illustrating Changes in Geographic Position.*

A further check is to be afforded through the observations for astronomic latitude at Ukiah. The observatory at Ukiah is between 25 and 30 miles in a direct line northeast of the fault. In connection with the general dislocation it was presumably moved toward the southeast and its latitude diminished by several hundredths of a second. This is one of an international series of observatories established in approximately the same latitude but in different longitudes, for the purpose of determining variations in the position of the earth's axis of rotation. If the observations at Ukiah were studied alone it might not be possible to separate the result of a small change in the observatory's position from the effects of the migration of the axis; but by combining the Ukiah data with those furnished by the other observatories of the system, it is probable that the effects of the two causes can be discriminated.

The most important practical results of the various earthquake studies will probably be afforded by the engineers and architects, and will lead to the construction of safer buildings in all parts of the country specially liable to earthquakes; but the geologic studies of the State Commission are not devoid of economic bearings. In the city of San Francisco and adjacent parts of the peninsula on which it stands the underlying formations include several distinct types, and the district is so generally occupied by buildings that the relations of the several formations to earthquake injury can readily be studied. Such a study is being made with care and thoroughness, and one of its results will be a map of the city showing the relation of the isoseismals, or lines marking grades of intensity, to tracts of solid rock, to tracts of dune sand in its natural position, to upland hollows partially filled by grading, and to old swamps, lagoons and tidal marshes that have been converted into dry land by extensive artificial deposits. The information contained in such a map should guide the reconstruction and future expansion of the city, not by determining the avoidance of unfavorable sites, but by showing in what areas exceptional precautions are needed, and what areas demand only ordinary precautions.

Another economic subject to which the commission may be expected to give attention is what might be called the earthquake outlook. Must the citizens of San Francisco and the bay district face the danger of experiencing within a few generations a shock equal to or even greater than the one to which they have just been subjected? Or have they earned by their recent calamity a long immunity from violent disturbance? If these questions could be answered in an authoritative way, or if a forecast could be made with a fair degree of probability, much good might result; and even if nothing more shall be possible than a cautious discussion of the data, I believe such a discussion should be undertaken and published. Of snap judgments there has been no lack, and the California press has catered to a natural desire of the commercial public for an optimistic view; but no opinion has yet been fortified by an adequate statement of the pertinent facts. Among these facts are the distribution of earthquake shocks as to locality, time and severity in California, and also in the well-studied earthquake district of Japan; the relation of the slipping that has just occurred to the geologic structure of the coast region; the relation of other fault lines to the bay district; and the relation of the recent shock to a destructive shock that occurred in 1868. If a broad and candid review of the facts shall give warrant for a forecast of practical immunity, the deep-rooted anxiety of the community will find therein a measure of relief. If a forecast of immunity shall not be warranted, the public should have the benefit of that information, to the end that it shall fully heed the counsel of those who maintain that the new city should be earthquake-proof. In any case, timidity will cause some to remove from the shaken district and will deter others who were contemplating immigration; but such considerations have only temporary influence and can not check in an important way the growth of the city. The destiny of San Francisco depends on the capacity and security of its harbor, on the wealth of the country behind it, and on its geographic relation to the commerce of the Pacific. Whatever the earthquake danger may be, it is a thing to be dealt with on the ground by skillful engineering, not avoided by flight; and the proper basis for all protective measures is the fullest possible information as to the extent and character of the danger.

Local Effects of the California Earthquake of 1906

By Stephen Taber
Stanford University

The principal damage done by the California earthquake of April 18, 1906, was confined to a long, narrow area extending along the Pacific coast in a northwest-and-southeast direction, with the city of San Francisco near its center. The area in which the greatest damage was done is a little over 200 miles in length and scarcely 40 miles in width. The earthquake may be accounted for by the geological structure. The principal valleys of California have been formed by a system of parallel faults running in a general northwest-and-southeast direction, and the disturbance occurred along one of these old fault-lines.

The particular fault which caused the earthquake is the Stevens Creek (Portolá-Tomales) fault; it has been traced across the Santa Cruz quadrangle by Dr. J.C. Branner and Dr. J.F. Newsom, and is described by them in the unpublished Santa Cruz folio of the United States Geological Survey. It runs from Crystal Springs Lake through Woodside and the Portolá Valley, over the saddle that joins Black Mountain with the crest of the Santa Cruz Range, down the Stevens Creek Cañon, crosses Campbell Creek about 2 miles southwest of Saratoga and continues in the same southeasterly direction toward Loma Prieta. From Crystal Springs Lake the fault has been traced toward the northwest by Professor A.C. Lawson through San Andreas Lake and out into the ocean near Mussel Rock, about 7 miles south of the Cliff House at San Francisco.

The topography indicates that the fault-line continues its northwesterly course through Bolinas Bay and Tomales Bay, and that it finally leaves the coast near Point Arena in Mendocino County.

The present paper deals only with the movements that took place along this fault-line between Crystal Springs Lake and Black Mountain, and with the effects of the earthquake on the district lying on both sides of the fault-line and extending from San Francisco Bay to the Pacific Ocean.

The Stevens Creek fault is one of the most recent, for it cuts gravel beds that were laid down as late as the Pliocene or perhaps Pleistocene period. The old uplift along the Stevens Creek fault is on the northeast side, and the rocks on both sides of the fault dip in a northeasterly direction. The Miocene sandstones that form the greater part of the Santa Cruz Range come down to the faultline on the west, but on the east side erosion has removed the overlying beds and exposed the Franciscan series, so that it is only at some distance away from the fault toward the east that the Miocene sandstones and gravels reappear.

At about fifteen minutes after five o'clock on the morning of April 18, 1906, the Stevens Creek fault was suddenly refractured, and a new displacement occurred along the old fault-line, producing the earthquake that shook the adjacent region.

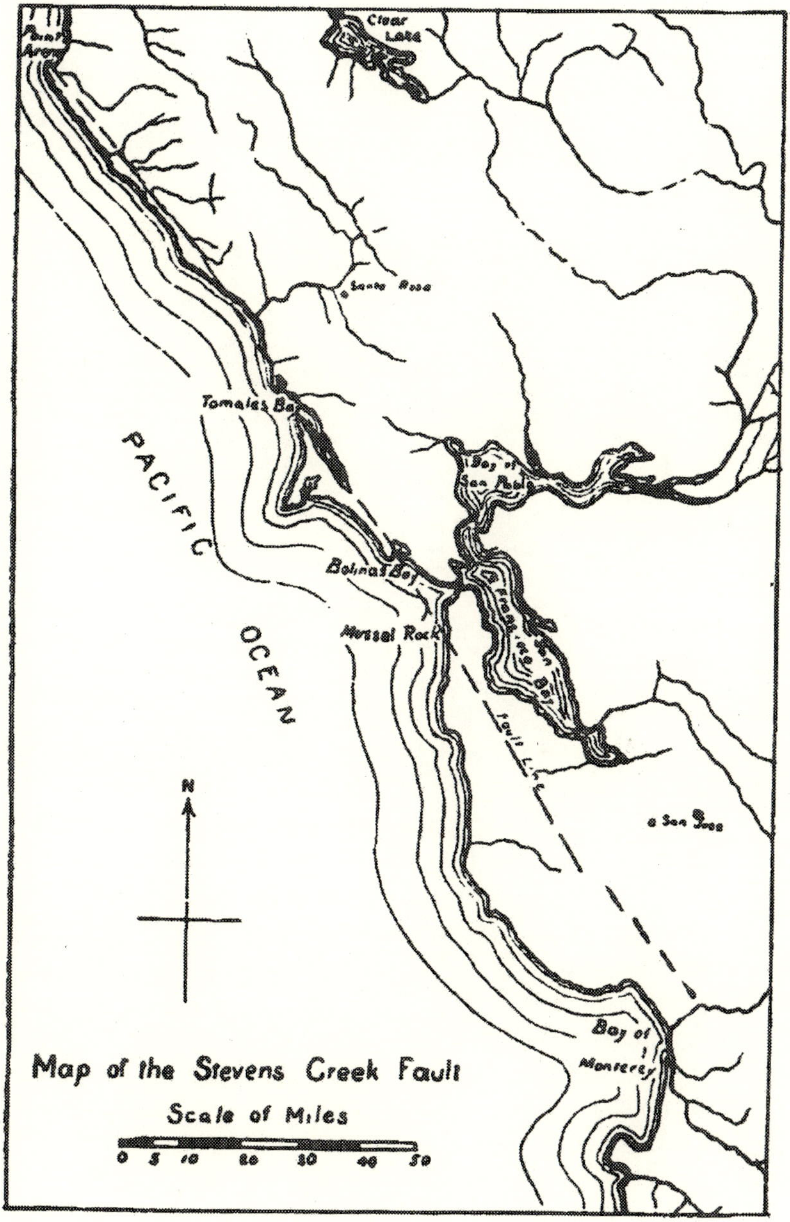

Clear Lake
Santa Rosa
Tomales Bay
PACIFIC
OCEAN
Bay of San Pablo
Bolinas Bay
Mussel Rock
Fault Line
San Jose
N
Bay of Monterey
Map of the Stevens Creek Fault
Scale of Miles
0 5 10 20 30 40 50

FIG. 1.

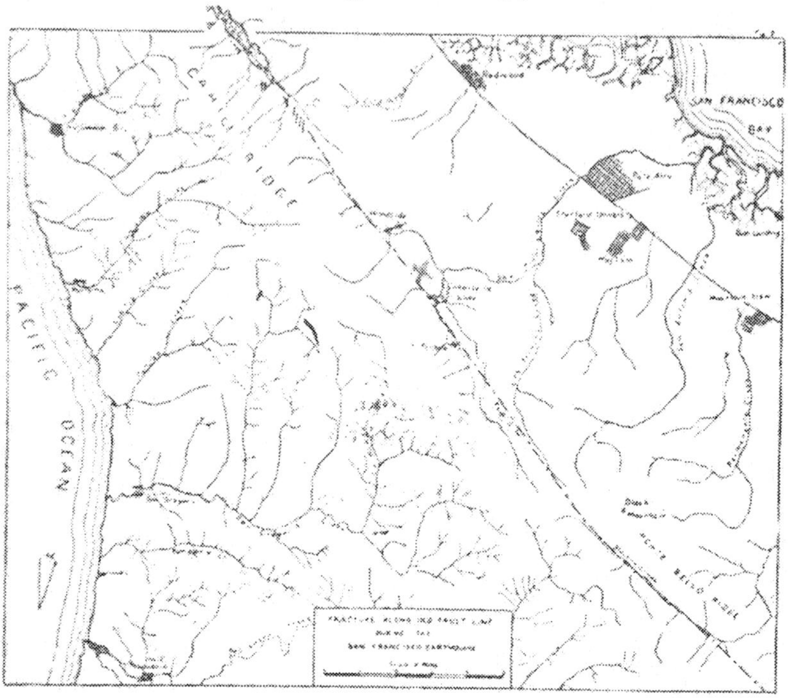

FIG. 2.

This new displacement is chiefly lateral, the southwest side of the fault having moved toward the northwest, or vice versa; and in some places this has been accompanied by a small uplift on the northeast, or a small downthrow on the southwest, or both. The lateral displacement is well defined, as far as it has been traced, and at some points amounts to as much as 9 feet. The vertical displacement in most places is not so evident, but about a mile southeast of Portolá there is an uplift of 2 feet on the northeast side, and the same amount of vertical displacement has been observed on Black Mountain.

The valleys through which the old Stevens Creek fault runs are filled with silt and gravels, so that it is impossible to get at bed-rock along the fault line, but it is probable that the rocks along this line

Fig. 3—Road Crossing the Fault Line Two Miles Southeast of Portolá. There Is a Vertical Uplift on the Northeast Side of the Fracture at This Place.

have been so broken and crushed by past movement that they would offer little or no additional information in regard to the recent displacement.

Through the Portolá Valley, and for about 3 miles northwest of Woodside, the fracture runs in a continuous and almost straight line. At a little distance it looks as though a furrow had been run down the valley with a big plow. In places the earth has been piled up into ridges 2 or 3 feet high, and at other places fissures have been opened that measure 2½ feet in width. Two and a half miles southeast of Portolá the fissure is 3 feet across.

The ground is usually cracked and broken for a distance of 10 or 15 feet on both sides of the main fracture, which in places splits up into numerous minor cracks.

FIG. 4—*Showing a Fence That Crossed the Fault at an Oblique Angle. The Post Shown in the Photograph Was Split and Pulled Apart and the Wires Broken.*

FIG. 5—*Showing a Fence That Was Broken and Offset Eight Feet Where It Crossed the Fracture.*

Fig. 6—*Photograph of a Fence Crossing the Fault-Line at Right Angles. The Man is Holding an Eight-Foot Transit Rod and Stands in Line with Continuation of the Fence on the Far Side of the Fracture. The Fence Was Repaired Before the Photograph Was Taken.*

Two miles southeast of Crystal Springs Lake the resistance to displacement appears to have been greater, and, instead of slipping along a straight line, the ground has been broken into a belt of parallel, north-and-south, shearing cracks, running at an angle of approximately 45° with the general movement. Some of these shearing cracks are from 1½ to 2 feet wide, and the belt of cracks extends for a quarter of a mile or more.

Black Mountain was badly shattered, and there are numerous cracks running over it in all directions. Fences crossing the fracture are broken; those that run in a north-and-south direction have their boards bent into arches or crushed, and the ends shoved past each other, while those that cross in a northwest-and-southeast direction have been pulled apart, and wire fences have been broken by tension. Fences that cross the fracture at right angles have been broken and displaced 8 or 9 feet.

The photograph (Fig. 6) shows a line fence crossing the fault a mile southeast of Woodside. This fence was broken and displaced over 8 feet, but had been repaired before the photograph was taken. The man at the right in the picture is holding an 8-foot transit rod, and he is standing in line with the continuation of the fence on the far side of the fracture. The crack crosses the fence just back of where he is standing.

A striking evidence of displacement is shown in the earth dam that divides the Crystal Springs Lake. This dam is about 500 feet in length, and the road from San Mateo to Half Moon Bay runs along its crest. The accompanying sketch (Fig. 7) shows the position and direction of the cracks that were formed in the dam. The larger cracks are about 6 inches wide and are parallel with the dam. Smaller intersecting cracks were formed near the northeast end of the dam along the probable line of the fault, and the road was offset about 6 feet at this point. The fences on both sides of the road were broken in a number of places, and the unbroken boards were bent and arched so as to give a serpentine appearance to the fences. The wires of a telephone line crossing the dam sag in great loops.

It seems probable that the total displacement is greater than the amount that may be directly measured at any place along the line of the fracture, for there is evidence of drag in the soil for a considerable distance on both sides. Water-pipes at a distance of several hundred feet from the fault-line have been pulled apart, telescoped, or bent in the direction of the movement, and fences formerly straight have been bent into a slight curve for a distance of 200 or 300 yards from the fracture.

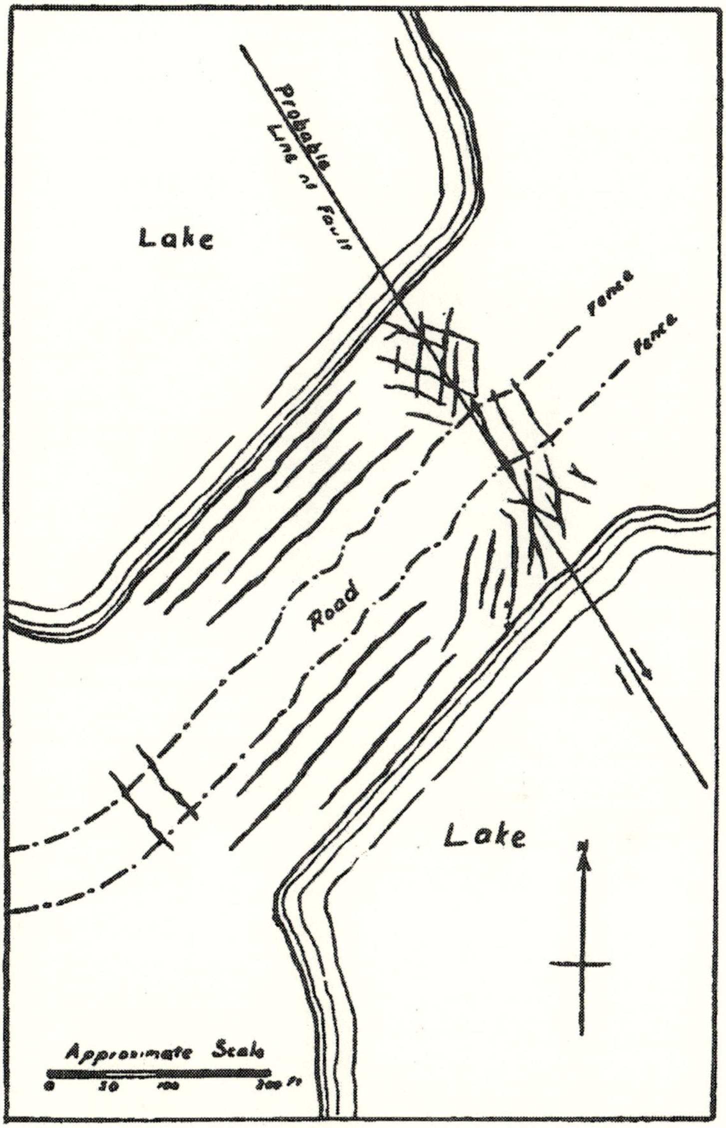

FIG. 7—*Dam at Crystal Springs Lake, Showing Cracks Formed by the Displacement.*

FIG. 8—*An Oak Tree Six Feet in Diameter Uprooted by the Earthquake Three Hundred Yards from the Fault Line.*

The intensity of the shock was greatest along the line of faulting and decreases as one goes away from this line. In comparing the intensity of the shock at different places, the best evidence was supplied by the oak trees broken and uprooted where the intensity was greatest, by the percentage of country water-tanks thrown down, and, as the intensity decreased, by the condition of plastering in houses and the number of brick chimneys found standing.

There are many white oaks (Quercus lobata) growing in the valleys through this section of the country, and in a belt extending for not more than 400 or 500 yards on each side of the fracture many of these trees have been uprooted or have had large branches whipped off. Sound limbs 2 feet thick were broken off by the shock, and there are trees having a diameter of more than 6 feet that were overturned during the earthquake. About 300 yards southwest of Searsville Lake a live oak (Quercus agrifolia), growing within a few feet of the fracture, was split down the trunk by the violence of the movement, but is still standing.

On Cahill Ridge, 2 miles southwest of the fault-line, there are redwoods (Sequoia sempervirens) that had their tops snapped off 75 or 100 feet above the ground. The intensity of the shock was much less at this point than near the line of fracture, but the redwood is brittle compared with the oak.

Frame houses, strongly built and having good foundations, stood the shock well, even when close to the fault-line, but brick and stone structures were badly damaged at distances of more than 12 miles from the fracture. Fortunately for the inhabitants, most of the houses near the fault-line were one-story frame buildings.

Most of the water-tanks that stood within 3 or 4 miles of the fracture were thrown down, but farther away the percentage of tanks that are standing gradually increases. With but few exceptions, all brick chimneys within 3 or 4 miles of the fault-line were thrown down, but at a distance of 8 or 9 miles probably more than 50 per cent are still standing.

Within the area under discussion the earthquake seems to have consisted of two separate and distinct kinds of movement: one a violent vibration in a northwest-and-southeast direction, parallel to the fracture, and probably caused by the sudden displacement; the other a wave-motion, traveling at right angles to the fracture and generated by the rocks slipping past each other along the fault-line.

It was the first motion that snapped off branches, overturned oak trees and wrecked buildings in the immediate vicinity of the fault-line; and although this motion extended for a considerable distance, the damage it caused was limited to a belt not over a mile distant from the fracture.

The following facts appear to bear out the theory of a violent initial movement parallel to the fault-line. Most of the trees that were overturned fell toward the northwest or southeast, and the buildings that were destroyed near the line of fracture tended to move in the same direction; but frame buildings do not furnish very reliable data. Beds and furniture rolled back and forth in directions parallel to the vibration. The strongest evidence is furnished by the movements of liquids, such as milk and water. In the immediate vicinity of the fracture many places were found where the water had splashed out of reservoirs and tanks on the northwest and southeast sides, and at one place the motion had been so violent that the water in a large wooden tank had splashed against a roof placed over it with sufficient force to drive shingles from the northwest side. In some places large water-tanks holding 3,000 or 4,000 gallons were almost emptied by the splashing.

The wave-motion was responsible for most of the damage done outside of the narrow belt along the line of fracture. Several people, who were out of doors at the time of the earthquake and several miles from the fault-line, state that the ground appeared to move like the waves of the sea. While these statements cannot be used as conclusive evidence, there are many facts that indicate a true wave-motion, having distinct crests. Water in reservoirs and tanks, standing at a distance of several miles from the fault-line, splashed out on the northeast and southwest sides. At King's Mountain House, a little over 2 miles southwest of the fracture, there were a number of milk-pans setting on shelves. All of the cream went out on the southwest side of the pans, and afterward the milk splashed back and forth, spilling out on both the southwest and northeast sides. At a barn 3 miles northeast of Woodside, heavy carriages standing with their wheels parallel to the fault-line were moved sideways a distance of 6 inches, but did not roll forward on their wheels.

At Stanford University, 4½ miles northeast of the fault-line, the sandstone buildings afford evidence of the wave-motion. Walls running northwest and southeast, when free to fall, fell by toppling over, and the stones lie on the ground in nearly the same relative positions that they occupied while standing. Walls running more nearly parallel to the direction of wave-motion were crushed, and the stones fell in irregular piles, while the walls that are still standing show 45° shearing cracks. Perhaps the best evidence of a true wave-motion is to be found in the arches. When the crest of a wave struck an arch running northeast and southwest, the arch was pulled apart, allowing the keystones to drop a short distance. There are forty-six arches running approximately northeast and southwest in which the keystones dropped, while only twelve arches running northwest and southeast had their keystones lowered, and some of the latter may be accounted for by the falling of neighboring walls. It might be well to state that there were more arches running northwest and southeast than at right angles to that direction. Most of the keystones dropped only 5 or 6 inches, but some fell out completely.

There are several strongly built, low, one-story frame houses, of the bungalow type, standing within a few hundred feet of the fault, which scarcely had their plaster cracked, excepting where chimneys fell through. Broken oak trees growing close to these houses indicate the intensity of the shock. This suggests that the wave-motion, with its shearing action, was more damaging to walls than a back-and-forth vibration. Another interesting fact in this connection is that most two-story frame buildings at a distance of 5 or 6 miles from the fault-line did not have the plaster cracked on the second floor, although the plaster on the first floor was usually badly cracked and broken.

Brick buildings at a distance of 10 miles from the fault-line showed the effects of a wave-motion. At Guth Landing, on San Francisco Bay, 9½ miles northeast of the fracture, there was a large brick warehouse, with its ends parallel to the fault-line. The upper half of each end toppled over, but the side walls, although badly cracked, were left standing.

FIG. 9—*Photograph of Arches at Stanford University, Showing Keystones Lowered During the Earthquake. These Arches Were Nearly at Right Angles to the Fault-Line.*

The effects of this wave-motion have not been traced more than 12 miles from the fault-line, but it probably continued with diminished intensity to a considerable distance. In other districts, having a different geological structure, the distances to which these movements could be traced would undoubtedly vary greatly. The wave-motion appears to have been more intense in the soft alluvial deposits of the valleys than in the consolidated beds that form the high ground, but there are not enough houses in the mountains of this district to furnish conclusive evidence on this point.

At Half Moon Bay the intensity of the earthquake was about the same as at Stanford University; but as one goes down the coast, and therefore away from the fault-line, the intensity decreases. At Pescadero, which is about 12 miles from the fault-line, there was scarcely any damage done, but there were no brick or stone buildings in that village.

In regard to the geological effects of the earthquake, there are a few facts of general interest that might be mentioned. Most of the landslides that occurred at this time were on the west side of the Santa Cruz Range. This is probably to be attributed to the greater rainfall on that side of the watershed. The springs and streams on both sides of the range increased in volume after the earthquake, and some creeks on the west side were nearly doubled. All of the streams were muddy for several days after the earthquake.

A marked effect was produced on the artesian belt near the head of San Francisco Bay. Wells that had previously been dry began flowing, and wells that flowed before the shock greatly increased in volume and pressure. The following is one illustration out of many that were recorded: A well near Alviso, at the head of the bay, formerly required a wind-mill to pump the water. At the time of the earthquake the casing was driven 2 feet out of the ground, wrecking the pump, and since that time the well has been flowing under a heavy pressure. In some of the lowlands small cracks formed, out of which water issued, bringing up mud and sand.

Preliminary Note on the Cause of the California Earthquake of 1906

By F. Omori, Sc.D.

Member of the Imperial Earthquake Investigation Committee

Tokyo, Japan

1. Introduction.

The great earthquake of April 18, 1906, which caused an enormous amount of damage in San Francisco, furnished a rare opportunity of studying the different earthquake phenomena, especially the seismic effects on various modern structures. Immediately upon receipt of the news of the catastrophe, the Imperial Government resolved to dispatch to California Professors T. Nakamura and T. Sano, and myself, for the purpose of making investigations on the great seismic disturbance, each according to his professional point of view. The party departed from Tokyo on May 1st, and arrived at San Francisco on the 18th of the same month, the present writer remaining about eighty days in California.

My special thanks are due to Professor George Davidson, and also to Professors Lawson and Leuschner of the University of California, Dr. Gilbert, of the U.S. Geological Survey, Mr. K. Uyeno, Japanese Consul, and other American and Japanese gentlemen, with whom I came in contact and who gave me most cordial assistance during my stay in California.

2. Time of Occurrence.

The times of earthquake occurrence observed at the California University and the Lick Observatory were respectively 5h 12m 39s and 5h 12m 12s A.M. (Western States Time, or that of longitude

120° W.); the time of commencement of the disturbance at the origin itself being probably about 5h 12m A.M.

3. Area of Destructive Motion.

The area, within which more or less damage was done, was very long, extending over a distance of 550 miles along the Pacific coast, from the vicinity of Salinas on the south to the vicinity of Eureka on the north. The width or extent from the coast of the strong motion area is probably some fifty miles. The earthquake of April 18th was thus greater, in length, than the large Japan earthquake of 1891, the length of whose area of strong motion was about 400 miles. The intensity of motion in the California earthquake was, however, less violent than in the other, and the amount of the casualties in San Francisco and the different parts of the strongly shaken zone was small comparatively.

4. Sea Waves.

When an earthquake of inland origin is large and violent, the waters of ponds, rivers or lakes are more or less disturbed. So similarly a great submarine earthquake is often followed by tidal waves; the time interval between the occurrence of the earthquake shock and the arrival of the destructive sea waves varies from a few minutes to several hours, and depends on the distance of the origin from the shore. Tidal waves which are not to be noticed on high seas are developed most markedly in bays with shallow waters and an open mouth, but are quite insignificant along deep-water straight coasts. Many of the great earthquakes originating off the Pacific coast of Alaska and Central and South America have been accompanied by large tidal waves. But fortunately, this phenomenon which sometimes causes more damage than the earthquake disturbances itself was so far not very destructive along the coast of the United States. The great earthquake of April 18th last produced distinct, but very small, disturbances of the bay waters which were clearly recorded on the tide gauge at the Presidio (San Francisco); the amount of the rise and fall of the sea water being only about 6 inches, repeated in about 40 minutes. Now the wave period or periods at a place on a given coast remain constant in all the tidal waves, irrespective of the origin or cause; a destructive tidal wave consisting simply in the increase of the amount of the water motion existing more or less at all times, in consequence of a strong submarine earthquake or eruption, a storm, or some other agency. A seismic tidal wave is caused by the movements communicated from the sea bottom to the superincumbent water mass: a very big water disturbance taking place when the earthquake focus is at the sea bottom itself or at a very small depth below it, accompanied by some changes in the contour of the sea bottom. The absence of any great tidal disturbance on April 18th shows that there was no great submarine depression or vertical dislocation, although it seems probable that the northern half of the epifocal zone was under the Pacific.

5. Sea Shock.

The steamer "Argo" felt the earthquake shock on sea near Cape Mendocino, the sensation being like that caused by running aground. There were other vessels which experienced the earthquake in a similar manner.

Effects like these, which may be called "seashocks," are due to the direct transmission through water of vibratory earthquake movements, and not due to the phenomena of the tidal waves which are developed only along coasts where there is some indentation.

6. Approximate Position of the Center of Epifocal Zone.

A rough idea as to the position of the most central or principal point in the zone, which forms the origin of the earthquake, may be obtained from a good seismograph record taken at the Lick Observatory, where the preliminary tremor lasted about 10 or 12 seconds, from which it may be calculated that the distance between the point in question and Mt. Hamilton was about 80 or 90 miles; the predominating direction of motion there being NNW and SSE. These data indicate a place near the Tomales Bay as the most central point of the disturbance. The approximate position of the latter may be assumed to be at a point, latitude 38°15'N, longitude 123°W.

PL. I.

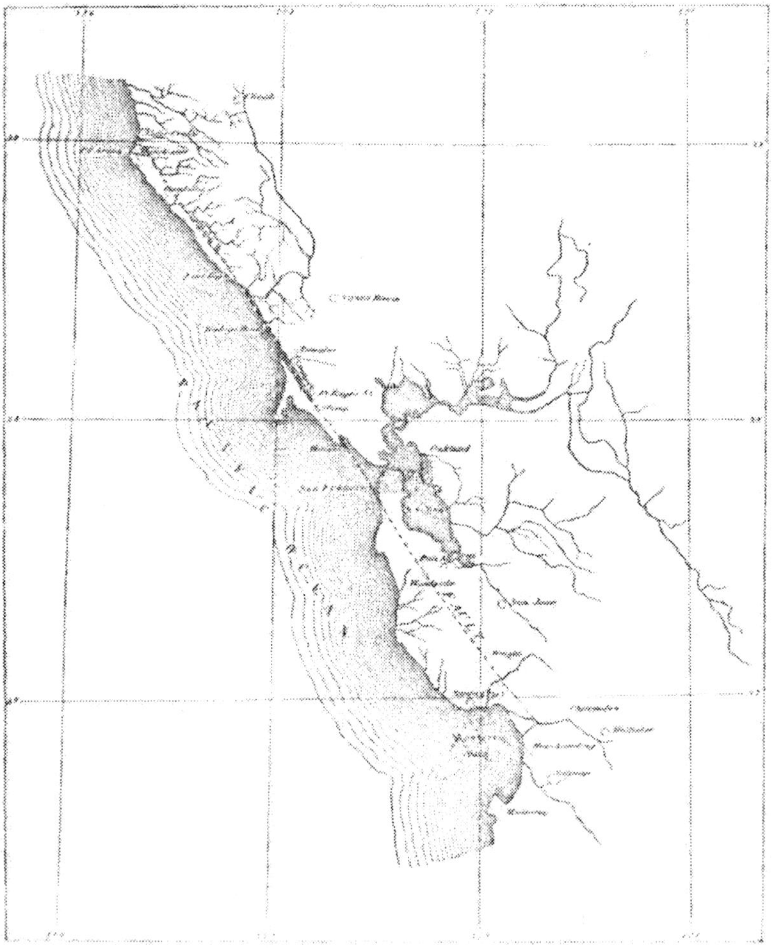

Fig. 1. San Francisco and the Vicinity. Showing the Course of the Great Fault, from Pt. Arena to Chittenden.

7. The Epifocal Zone.

One of the peculiar features in the topography of the State of California is a straight depression whose direction is NNW and SSE and which extends through the valley of the Gualala River, and Tomales and Bolinas Bays, continued further southeastwards for some distance. This depression, which must have been formed in bygone ages by a great sudden convulsion of the earth's crust, or by the gradual mountain-making force going along the Pacific Coast, shows signs of dislocations caused at no very remote epoch by some great earthquakes, and it is of a special interest that the earthquake of April 18th again produced along the same old weak zone a continuous series of remarkable surface manifestations of cracks, depression, or horizontal slipping, constituting what is called a "fault" in geology. This fault which has been most carefully studied by Dr. Gilbert of the U.S. Geological Survey, Professors Lawson and Branner, and other able geologists of the California and Stanford Universities, begins on the north at the right-hand mouth of the Alder Creek, near Pt. Arena, and passes into the ocean at the vicinity of Fort Ross; it again appears at the Bodega Head and at the eastern side of the mouth of Tomales Bay, crosses to Inverness, on the west shore of the same bay, and then passes through the vicinity of Pt. Reyes Station, continued to a place about 4 miles to the west of the Stanford University; marked disturbances of the ground being also distinctly shown to the southeast, in the vicinity of Wrights and Chittenden. The length of the visible fault is thus over 150 miles, being three times that of the fault line in the great Japan earthquake of 1891. It is, further, extremely probable that the northwestern part of the present fault is continued beyond Pt. Arena under the ocean some 120 miles more and extends to the vicinity of Cape Fortuna. That the fault was not a mere surface phenomenon is shown by the appearance of the same disturbance across the tunnel near Wrights Station, at a depth of some 700 feet below the mountain surface. See Fig. 1 (Pl. I) and Fig. 2 (Pl. II).

8. Shear of the Ground.

The shearing movement of the ground produced many remarkable results; roads, fences, and every other thing crossed by the line of disturbance being cut apart and displaced considerably. There were cases in which even large redwood trees were split by the shearing motion of the ground.

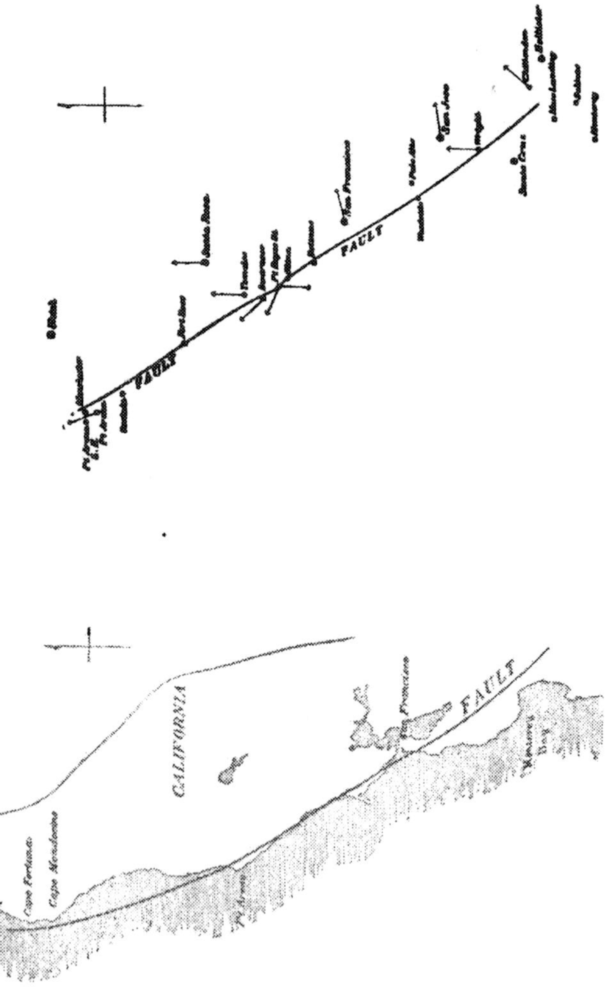

Pl. II.

Fig. 8. Directions of Motion at Different places on or near the Great Fault.

Fig. 9. San Francisco and the Vicinity, showing the General Course of the Great Fault.

Fig. 4 relates to the shear effect observed near Olima, a village situated between the Tomales and Bolinas Bays. The fault passed just in front of a house (Skinner's Ranch) and produced a relative displacement of 16 feet, a garden walk being carried through that distance from a to b.

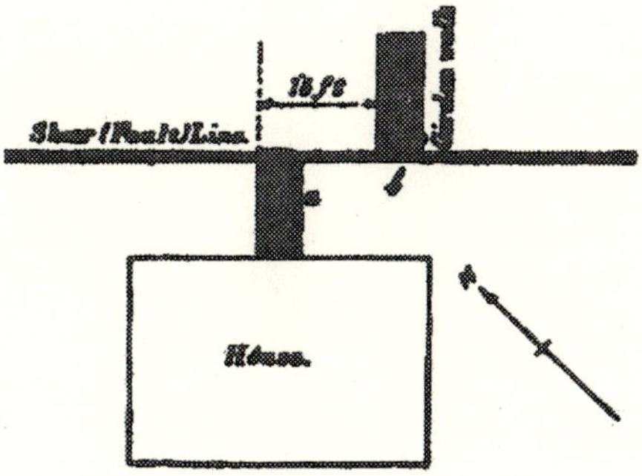

Fig. 4.

Fig. 11 (Pl. III) shows the shearing effects on a pier at Inverness, on the west coast of the Tomales Bay. The end part of the pier was separated from the rest and was displaced about 20 feet toward NNW. The direction of displacement in this particular instance was opposite to the general direction of the relative slip along the great fault line.

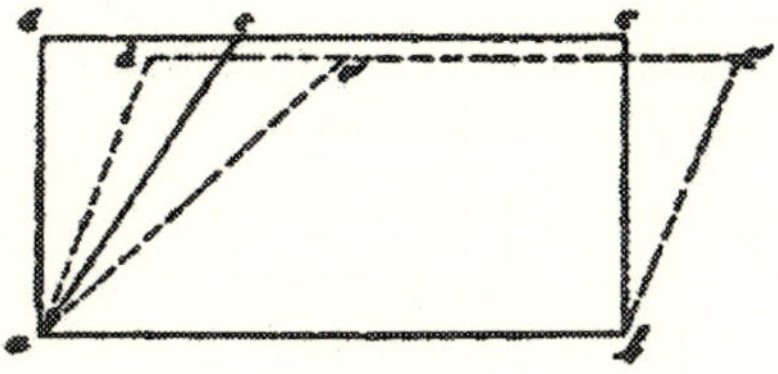

Fig. 5

Fig. 12 (Pl. III) shows one of the fault cracks produced among the hills above Fort Ross. It will be observed that the new disturbances appeared along a depression marked by a series of small ponds (shown at the right-hand side of the cut), these latter being traces left by a former great earthquake.

Fig. 13 shows the remarkable compression and shear effects along one of the parallel fault cracks, observed on elevated grounds near the town of Manchester, not far from Pt. Arena. A foot-scale placed in the foreground will show the size of the overlapping earth pieces, whose plan is given in Fig. 8.

9. Remarks on Shearing Movements.

For the sake of illustration, let us first consider cracks of a wall when the earthquake motion is parallel to the latter.

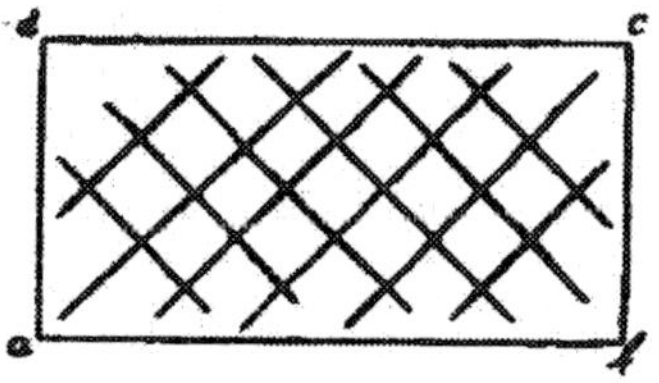

Fig. 6.

Let a b c d (Fig. 5) be a wall whose bottom side a b is fixed, either absolutely or relatively, while the upper side c d is brought to the position c' d' as the result of a shearing stress in the direction of a to b. Then the rate of the length change of the line a e, connecting a with any point e on the side c d, will be greatest when the angle d a e is equal to 45°. Consequently there will be formed a series of cracks at right angles to the lines of greatest elongation and at an angle of 45° to the base a b.

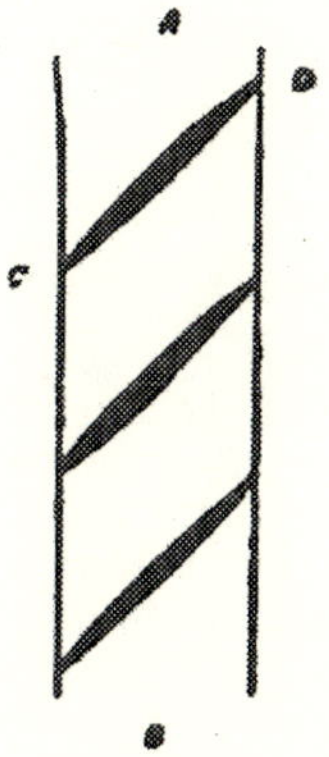

Fig. 7

AB... Fault Line.
CD... Shear Cracks.

Thus, in the case of a strong horizontal motion parallel to the plane of the wall, there will be two sets of cracks at right angles to one another, as in Fig. 6.

Fig. 14 (P1. IV) illustrates some of the cracks of plastered walls observed in St. James Hotel, San José.

Secondary Cracks of the Ground.—Along the fault line the ground was, as in other cases, very often bulged up, forming a narrow zone of 1 or 2 feet elevation and some 5 or 10 feet width, as if raised up by a gigantic mole creeping underground. This sort of ridge, whose formation was due to the shearing action, combined with a compression along the line of dislocation, showed usually a series of secondary oblique cracks, as is diagrammatically indicated in Fig. 7. These ground cracks were perfectly similar to the shear cracks of walls considered above.

Figs. 8, 9 and 10 show parts of the fault lines found near the town of Manchester, not far from Pt. Arena; the dotted lines in each figure indicating the directions of the secondary shear cracks. Fig. 8 is the plan of the remarkable disturbances shown in Fig. 13. In Figs.

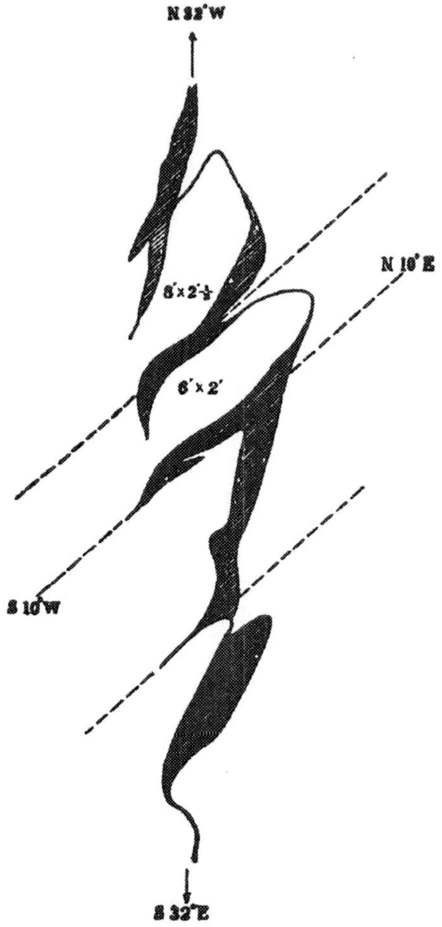

Fig. 8—*The Shaded Parts Indicate Cracks of the Ground.*

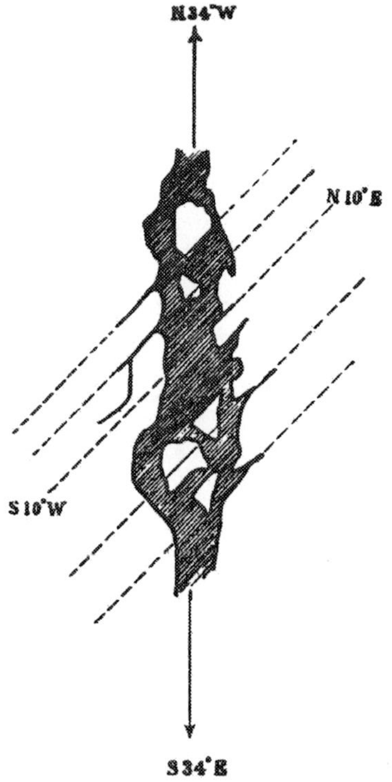

Fig. 9.

9 and 10, the angle between the main fault line and the shear cracks varied between 44° and 47°. In Fig. 8, however, there was evidently a very strong compression, and the shear angle was smaller, namely, 42°.

I have measured the shear angle in eleven other cases, where it varied between 35° and 47°; the total average value being 40°.

If the shear be accompanied by a horizontal compression at right angles to the fault line, the angle between the latter and the shear cracks will be smaller than 45°, as suggested by Professor A. Inokuty, of the Engineering College, Tokyo Imperial University. The coexistence of a tension normal to the fault plane will, on the other hand, make the same angle greater than 45°.

10. Comparison with the Formosa Earthquake of March 17, 1906.

The local but very severe earthquake in the Kagi Prefecture, Formosa, on March 17, 1906, produced also remarkable surface dislocations, in which the vertical depression and the horizontal shear each amounted to about 8 feet. The angle between the direction of the main fault and that of the shear cracks was on the average 43°.

11. Landslips, etc.

In the meisoseismal area, there were great many cases of mountain slides. The most remarkable among these was that which occurred near Cape Fortuna (False Cape), where an enormous quantity of debris was detached from a mountain side and was pushed into the ocean, creating a new promontory of about 3/4 mile length.

At Moss Landing, near Salinas, there were great horizontal disturbances of the sandy ground; the office of the station agent being displaced about 15 feet relative to the adjoining fence.

12. Direction of Motion in San Francisco.

Fig. 19 (P1. VII) shows the directions towards which 520 monuments at the different cemeteries in San Francisco and the vicinity were overturned by the earthquake shock. It will be observed that the greatest number of the monuments were overturned towards the east or east slightly north. The mean direction of overturning is N76°E, which may be regarded as the direction toward which the greatest horizontal displacement took place.

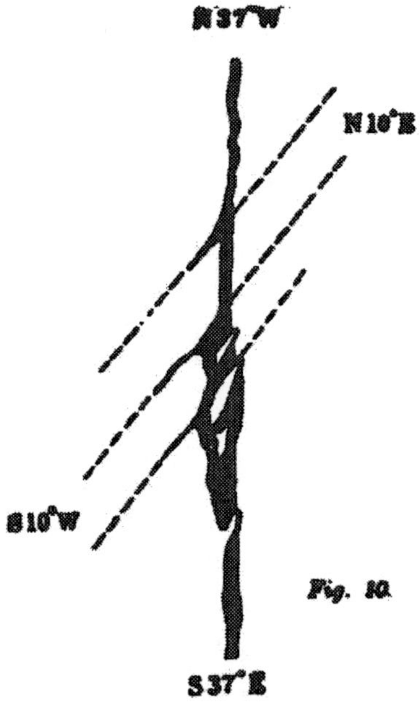

FIG. 10.

13. Relation to the Great Fault of the Directions of Motion at the Neighboring Places.

The approximate directions of the principal or strongest motion at the different places on or near the fault, each determined from numerous overturned bodies, were as follows:—

A.

A San Francisco: N76°E.

San Jose: N81°E.

Chittenden: N38°E.

Watsonville: NE.

Santa Rosa: N.

Tomales: N.

Pt. Reyes Station (East side of Fault): S.

B.

B. Pt. Arena: NNW.

Inverness: NW.

Pt. Reyes Station (West side of Fault): WWN.

Wrights: N

The mean general direction of the fault is N 37° W—S37°E, this being exactly identical with the direction of the great depression zone before mentioned. The places in Group A are situated on the eastern side of the fault line, while those in Group B are situated on the western side. It will thus be observed that at the A Group places the direction of motion was mostly towards north-northeast, or northeast by east; while at the B Group places, the direction was toward northwest, north, or northwest by west. Thus, on the whole, the motion on each side of the fault line had a tendency to diverge, or to be directed away, from the latter. This can be explained on the supposition of a subterranean collapse, or settling down, which would produce an initial inward motion, to be followed by the second and larger outward displacement.

PL. III.

FIG. 11—*The Shearing Effects on a Pier at Inverness, on the West Coast of the Tomales Bay. The End Part of the Pier Was Displaced About 20 Feet Towards NNW.*

FIG. 12—*One of the Fault Cracks Produced Among the Hills Above Fort Ross. The New Disturbances Appeared Along a Depression Marked by a Series of Small Ponds (Shown at the Right-Hand Side of the Picture), Which Are Traces Left by a Former Earthquake.*

Further, the directions of motion at the different places were mostly northward, and not southward. This would mean that the whole meisoseismal zone was first pushed towards SSE, the second or counter motion, which was greater, being consequently directed toward NNW. I presume, therefore, that the action which caused the great earthquake of April 18th was a sudden movement towards southeast by south of the earth's crust at the west coast of California, accompanied by some downward thrust. In this connection it is extremely interesting to note that Mount Tamalpais, on the north shore of the Golden Gate, has been ascertained from trigonometrical measurements, to have moved, between 1851 and 1882, 5.6 feet towards N12°W, indicating that the earth's crust at this part of America was being strained toward the same direction. The ground on the eastern side of the fault line was generally displaced toward SSE relative to the ground on the other side; the amount of the horizontal slip was maximum at places between Pt. Arena and Pt. Reyes Station and varied from 16 ft. to 20 ft.; the amount of displacement decreasing to about 8 ft. at Woodside, near Stanford University, and to about 4 ft. in the vicinity of Wrights. From the uniformity of northward direction of motion it is probable that both sides of the fault line were displaced toward NNW, but the west side was moved more than the east side, the amount of the horizontal slip, or shear, above mentioned, being merely relative or differential. In the majority of cases the eastern side was depressed, the maximum amount being 2 feet.

14. Depth of the Disturbance.

From the comparatively very small number of after-shocks, I am inclined to suppose that the main source of the earthquake was situated some considerable depth below the surface. In fact, the earthquake seems to have been caused by a disturbance which took place along the old weak line, but extended deeper into the earth's crust. The great depth of the main source of disturbance also

PL. IV.

FIG. 13—*Remarkable Compression and Shear Effects Along One of the Fault Cracks, Produced on Elevated Grounds Near Pt. Arena. A Foot-Scale Placed in the Foreground Shows the Size of the Overlapping Earth Pieces, Whose Plan Is Given in Fig. 8.*

FIG. 14—*Some of the Cracks of Plastered Walls, in St. James Hotel, San José.*

explains why the intensity of motion was comparatively not very violent, and also why some places, such as Santa Rosa, San José and Ferndale, not directly on the fault zone, were also badly shaken.

15. Earthquake Damage.

This earthquake enabled us, for the first time, to study the effects of the shocks on steel-brick and reinforced concrete buildings; there being also numerous other damaged structures, such as ordinary brick, stone and wooden houses, bridges, water-pipes, etc. In San Francisco the earthquake was followed by fires, which broke out from several places, continued for three days, and entirely destroyed the principal business quarters of the city. The total area of the burned districts was 4.1 square miles, which is equivalent to 6 times the area of the great London fire of 1666. The amount of casualties was, however, comparatively small, the ascertained number of persons killed being 390. The total number of the killed in the whole earthquake area was probably not more than 1000, the loss of life in Santa Rosa, Stanford University, and other strongly shaken places being slight. In San Francisco serious damage was confined to the filled-up grounds, where the motion was not so strong as in the cities of Nagoya (max. acceleration=2600 mm. per sec, per sec.), Fukui (max. acceleration=2500 mm. per sec. per sec.), etc., on the occasion of the great Mino-Owari earthquake of 1891. The double amplitude of motion in San Francisco was probably some 4 inches, and the complete period of vibration about 1 second.

Fig. 15 shows the damaged condition of the newly erected Library of the Stanford University. The central steel dome, which is virtually an elastic inverted pendulum, evidently much vibrated, thereby causing destruction to loosely connected brick and stone parts of the building. The mortar used for cementing the masonry walls was of an exceptionally bad quality.

The damage to the City Hall of San Francisco was also principally due to the same two circumstances, stances, namely, the vibration of its high steel tower, and the bad quality of mortar.

PL. V.

FIG. 15—*The Damaged Condition of the Newly Erected Library of the Stanford University. The Central Steel Dome Behaved as an Elastic Inverted Pendulum.*

FIG. 16—*The Ruined Condition of a Steel-Framed Brick House in San Francisco, Which Was Dynamited and Then Burnt, Showing the Remarkable Effects of the Intense Heat.*

Fig. 16 (P1. V) shows the ruined condition of a steel-framed brick house in San Francisco, which was dynamited and then burned. The effect of the intense heat is remarkable, the steel frames being distorted in every possible form, as if they had been formed of a soft malleable metal.

Fig. 17 (P1. VI) gives an outside view of the back part of the Observatory on the top of Strawberry Hill, in the Golden Gate Park, San Francisco. This building is of reinforced concrete and furnishes a good demonstration of the strength of such structures. The Observatory was indeed seriously damaged and its front portion fell down to the ground, but this was on account of the weakness of the foundation ground, which was mostly a filled-up one and was considerably cracked and depressed. Fig. 18 shows, in a larger scale, one of the cracks of the basement wall, similar to that shown in Fig. 17. The steel cable, one inch in diameter, which was embedded in the concrete, was broken. The use of steel cables in concrete walls thus seems to be objectionable, as they are more liable to rusting than solid steel bars. None of the latter used in the concrete walls and floor of the Observatory, whose section was half inch square, was found broken; the adhesion of concrete and steel being also very good.

16. Recent Seismic Activity.

Recently there have been a number of great earthquakes in different parts of the world, especially along the following two zones:—

(A). The Pacific coast of North and South America.

(B). Himalayas and North Mediterranean zone. Next two sections give a short account of the earthquakes belonging to these two zones.

PL. VI.

FIG. 17—*The Observatory on the Top of the Strawberry Hill, in the Golden Gate Park, San Francisco, Built of Reinforced Concrete. An Outside View of the Back Part.*

FIG. 18—*One of the Cracks of the Basement Wall. The Steel Cable, One Inch in Diameter, Which Was Embedded in the Concrete, Was Broken.*

17. Earthquakes along the West Coast of North and South America.

Within the 7 years preceding the California earthquake of April 18, 1906, there were, along the Pacific coast of the American continents, seven great earthquakes, on the dates as follows:—

(i) Sept. 4 and 11, 1899; and Oct. 9, 1900.

(ii) Jan. 20, 1900; and April 19 and Sept 23, 1902.

(iii) Jan. 31, 1906.

Fig. 20.

Of the above seven earthquakes, the three of group (i) took place off the southwest coast of Alaska, two of them being accompanied by great tidal waves. The three earthquakes of the group (ii) strongly shook Mexico and Guatemala (Central America); while the earthquake of group (iii), which was accompanied by tidal disturbances, caused considerable damage in Panama, and the west coast of Columbia and Ecuador. The approximate positions of these three groups of earthquakes are marked in Fig. 20 by dotted lines, 1, 2, and 3.

As the west coast of the American continents is one of the great seismic zones on the earth, it is to be supposed that the seven destructive earthquakes above enumerated were not separate or local phenomena, but were the results of great stresses going on along the Pacific coast zone, extending from Alaska to South America, manifested at its north and middle parts. Hence an event most naturally to be expected would have been the extension of the seismic disturbance to the west coast of the United States, which so far had been free from the visitation of disastrous earthquakes. This apprehended event finally took place on April 18, 1907, the approximate position of the origin being indicated in Fig. 20 by a thick line marked 4. The great California earthquake may, therefore, be regarded as having completed the continuity of the seismic activity along these districts, which latter thus become, for a certain number of years, say 20 or 30 years, seismically a very safe place; large earthquakes, which remove a great unstability in the earth's crust, never happening successively at once and the same place.

During my recent stay in San Francisco I explained on several occasions reasonings like the above to newspaper reporters and others, also pointing out that even in the case of a future destructive earthquake, the intensity of motion would not be extremely violent, so that a slight amount of precaution taken in building houses would ensure an almost perfect immunity from earthquake shocks. As to the probable position of the next great shock on the Pacific side of America, I expressed my view that it would be to the south of the equator (that is to say, Chile and Peru), This is what I published in the San Francisco Bulletin of June 13, 1906. as it was very likely that the seismic activity would extend to either end along the great zone in question, and as the coasts of the countries above named are often visited by strong earth convulsions. I departed on

August 4th from San Francisco for home, and arrived on the 22d of the same month at Yokohama, first there learning of the disastrous shock of Valparaiso, which confirmed my anticipation. The approximate position of the origin of this last earthquake, which took place on August 17, 1906, is indicated in Fig. 20 by a thick line marked 5.

The great stresses going on along the whole Pacific coast of America, which thus resulted in the occurrence of a series of great earthquakes, seem to be connected with the growth of the Rocky and Andes mountain ranges; the Valparaiso earthquake bringing probably the great seismic activity along the zone under consideration for the time to an end.

18. Activity along the Himalayas and North Mediterranean Zone.

With regard to the seismic activity in Asia and Europe, it is to be noted that the unusually severe eruptions of Vesuvius, which began on about April 7, 1906, lasted about one week, and ended on the 13th of the same month. On the following day, namely April 14th, there took place the destructive earthquake of Kagi District (Formosa), in which 1249 persons were killed. Four days later there took place the great California earthquake. Whether or not there existed a connection between the Vesuvian eruption and these earthquakes, it is a matter of fact that there was a great seismic activity along the whole length of the zone extending from the north coast of the Mediterranean to the Himalayas, and possibly to Formosa. The different earthquakes belonging to the zone in question, which happened recently, are as follows:

(i) Assam and Bengal (India), June 12, 1897.

(ii) Aidin (Smyrna), Sept. 20, 1899.

(iii) Schemacha (Caucasus), Feb. 13, 1902.

(iv) Kashugar (Turkestan), Aug. 22, 1902.

(v) Saloniki (Macedonia), April 4, 1904.

(vi) Kagi (Formosa), April 24, 1904.

(vii) Kag (Formosa), Nov. 6, 1904.

(viii) Kangra Valley (the Punjab, India), April 4, 1905.

(ix) Calabria (Italy), Sept. 8, 1905.

(x) Kagi (Formosa), March 17, 1906.

(xi) Kagi (Formosa), April 14, 1906.

Thus great earthquakes took place at the different parts of the zone stretching through Italy, Macedonia, Asia Minor, Caucasus, Turkestan, the outer side of the Himalayas, and Formosa; this proving that the underground stresses were growing along the whole zone. As the seismic disturbances above enumerated occurred in the same epoch as those belonging to the American zone, it is extremely likely that underground stresses reached a maximum all over the earth, resulting in a marked display of seismic disturbances along certain zones of weakness.

19. Conclusion.

Future studies in various phenomena connected with the movements of the earth's crust might perhaps tend to advance our knowledge respecting the problem of the prediction of great earthquakes, which are often preceded by what may be called "fore-shocks." In the meanwhile, and always, it will be necessary to build houses and other structures strong enough to resist earthquake shocks, a problem which presents no great difficulties.

Tokyo, Nov. 1, 1906.

The Great Earthquake Rift of California

By Harold W. Fairbanks, Ph.D.

Reprinted from the *Bulletin of the California Physical Geography Club,* October, 1907

Whichever way we turn we are constantly impressed by the fact that the surface of the earth, with all its varied features, is undergoing continual change. The forces without, tearing down the mountains and carrying their materials to the lowlands, as a usual thing act slowly and quietly; but the forces within are frequently violent as they build up lofty volcanoes or break the crust and raise it in precipitous blocks.

Although we dread earthquakes with all their resultant destruction, yet it is well to recognize the fact that if it were not for them we would find here in California little of that wonderful scenery of which we are so proud. Our earthquakes are due to movements similar to those which, through hundreds of thousands of years, have been raising the lofty mountains of the Cordilleran region. The Sierra Nevada range, with its abrupt eastern scarp nearly two miles high, faces an important line of fracture along which movements have continued to take place up to the present time. The Owens Valley earthquake of 1872 was the most recent on this line and gave rise to a vertical displacement of 10 to 40 feet.

The presence of another line of weakness in the earth's crust was brought forcibly to our attention in April, 1906, and, although it has not yet resulted in the formation of a continuous line of lofty mountains, it may eventually do so. The great earthquake rift of California is the most remarkable thing of its kind known in the world. This rift or fracture in the crust of the earth extends in a northwest and southeast direction for 700 miles across California. The rift lies for the most part in the Coast Ranges, but passes into the ocean upon the north and the Colorado desert upon the south.

The rift must not be conceived of as an open fissure, for the walls are tight. The earth has broken under stress, and the walls upon opposite sides of the break slipped upon each other until the strain was in large part relieved. The earth movements have been of a complex nature, so that, while in places no vertical displacement appears to have taken place, in others it has produced escarpments 300 to 400 feet high. The walls of the fissure not being absolutely even and straight, they would grind upon each other where two convex surfaces came together, producing a ridge upon the surface of the earth. Where two concave surfaces came opposite each other, the earth would cave in, giving rise to hollows upon the surface. Thus we have produced along the rift peculiar features not due to the ordinary forces of erosion, features constituting what we might call "earthquake topography."

Lake in a Sunken Area on the Rift in Cholame Valley.

The clay formed by the grinding of the walls cuts off underground water-courses, forcing the water to the surface and forming springs, which along the more arid portions of the rift are among the most characteristic features. In the hollows due to the sunken earth the water collects and gives rise to ponds and lakes. The presence of these peculiar features enables us to trace the rift throughout its whole course.

As a matter of fact, certain portions of the great rift have been known to the country people living along it for many years. Especially is this true of the southern portion, which opened in 1857, and that lying in San Benito County, which has opened several times since that date. Previous to the earthquake of April, 1906, the writer had traced the rift for fully 400 miles. When the last earthquake came, the northern portion of the rift then known opened again, and, in addition, directed our attention to nearly three hundred miles more of its course which had not previously been explored.

A remarkable thing in connection with the recent earthquake was the horizontal movement of the walls of the fissure. In most fault fissures the movement has been vertical or oblique, as is shown by the striations upon the walls, but in the present instance there appears to have been no general vertical component. Wherever scarps appear they are local and due either to the settling of the earth upon a hillside or to the sagging of the earth where the movement brought two concave surfaces together.

The eastern wall of the great rift moved south and the western wall moved north a distance diminishing from 16 feet near Point Arena to 4 feet south of the Pájaro Cañon, San Benito County, and to nothing near the town of San Juan. It is clear that less than half of the known rift opened during the recent disturbance. It is probable from what can be learned that the whole southern portion of the rift opened in the earthquake of 1857, while at least two local openings have occurred in the middle portion in San Benito County between the dates of the two great earthquakes referred to.

Effect of Recent Earthquake in the Ridge Above Mussel Rock.

We have no idea of the date of the original opening of the rift. The so-called "earthquake topography" by which we trace the rift had its inception hundreds of years ago, for huge oaks are growing upon it in places. The origin of the fault fissure was undoubtedly far back in the geological history of the region. Along many portions of the rift there are faults showing a vertical displacement of more than 1000 feet, the effects of which do not appear at all in the present topography.

The crushed and broken rocks along the rift are more easily eroded than solid rocks, and this has given rise to cañons and valleys, which for many miles mark certain portions. These features have determined the location of roads and trails. The springs also have determined the position of houses and ranch buildings. Thus for years the rift features have had their influence upon people without the latter recognizing their meaning and importance. The effect of the rift upon the water supply is most noticeable through the Mohave Desert and San Bernardino Valley. The rift extends for many miles along the northern base of the Sierra Madre in the edge of the Mohave Desert. The underground waters seeping down the valleys are forced to the surface by the impervious clays, and thus give rise to valuable springs, cienegas and ponds. These become of extreme value in a region where the lack of water is the only bar to settlement.

The northernmost point where the rift has been observed is upon the projecting headland near Shelter Cove, southern Humboldt County. Thence southward for some fifty miles it lies in the ocean a short distance off-shore. The rift comes inland again at the mouth of Alder Creek near Point Arena. The valleys of the Garcia and Wallala mark its southward course, and two miles southerly from Fort Ross it enters the ocean again. Passing across Bodega Head, the course of the rift carries it the whole length of Tomales Bay, then through the valley-like depression leading over to Bolinas Bay.

Road Displaced 12 Feet on Line of Rift at Head of Tomales Bay.

The rift lies about a mile off-shore opposite the Golden Gate, but at Mussel Rock, six miles southward, it encounters the land, and from this point, as we follow its course, it carries us farther and farther inland.

From Mussel Rock to San Andreas Lake the rift is marked by a series of ponds and small lakes. Back of Stanford University the rift traverses the Portolá Valley, and there it is known as the "Portolá fault." The course pursued finally takes it over the Santa Cruz Range and down to the Pájaro Cañon at the point where the latter is spanned by the Southern Pacific Railroad bridge. The horizontal movement was here about 4 feet, and it is interesting to note that the rift opened between the western abutment and the adjoining pier, increasing the distance between the two so much that the iron girders nearly dropped into the river.

The town of San Juan, San Benito County, is situated most peculiarly with reference to the rift line, and, although many farm buildings owe their position to conditions of the rift, this is the only town whose location has been determined by it. The town lies upon a slight eminence in the broad valley of the San Benito River. A detailed examination shows that this eminence slopes gently to the southwest, while upon the northeast there is a steep and regular bluff fifty feet high. The existence of springs at the base of the bluff, and the fact that it lies directly in the line of the rift, as well as the general resemblance of the eminence to known fault blocks, make it reasonable to assume that such is its origin.

From San Juan we trace the rift on its regular southerly course along the northern slope of the Gavilan Range. Near the head of San Juan Cañon movements along the rift have led the stream which occupied the cañon to break over a low spot in the watershed and, as a consequence, it abandoned the lower part of the cañon and now flows directly down the mountain toward Hollister.

Alkali Sink on Line of the Rift North of Bakersfield.

A few miles farther on we reach Cienega Valley, where the uplift of the northern wall of the rift has made a dam across a valley. Back of this dam, gravel has collected, forming an underground reservoir, from which the town of Hollister gets its supply.

Earth movements have made the rift in this section the most suitable place for a road, therefore we follow it most of the way to the San Benito River and then up the valley of this stream for many miles. Shortly after reaching the river we come to a remarkable depression, nearly a mile long, due to subsidence during some one of the violent disturbances of long ago. By means of ridges, hollows, and ponds, the rift is traced in a direct line to Bitterwater, Monterey County.

South of San Benito Postoffice appears a fault block similar to, although much larger than, the one at San Juan. This fault block has an abrupt eastern slope, two hundred feet high, along the side of which the road runs. It has a long, gentle slope toward the west. This fault block is a good representation in miniature of the great Sierra Nevada fault block.

Next on the line which we are following lies Dry Lake Valley, comprising an area of considerable extent, which has been so disturbed by earth movements that it has no external drainage. Looking southeasterly from this valley we see a precipitous mountain profile in the distance and directly on the line of the rift. Upon examination this appears to be the product of some earthquake movement. Across one small valley the road follows the top of a ridge which has a striking resemblance to an old railroad embankment.

Bitterwater Valley is partly occupied by a lake during the wet season as a result of earth disturbances. Where the rift crosses Lewis Creek there is an enormous landslide which nearly blocks the valley. Along the mountain ridge north of Peach Tree Valley there are many landslides, and in Stone Cañon rapidly eroding slopes testify to comparatively recent movements.

The rift traverses Cholame Valley in eastern Monterey County. Near Parkfield the surface over a considerable area bears evidence of having been greatly broken up. A branch rift with two parts, one marked by the front of a line of hills and the other by depressions and lakes, is traceable for several miles in a more easterly course. West of Parkfield a giant oak, undoubtedly many hundreds of years old, stands upon a low ridge which marks the main rift. At this place there is a spring and a pretty home, the one due to, the other made possible by, the existence of the earth fracture.

From the lower end of Cholame Valley the rift passes over rolling hills until it reaches the Carrisa Plain, down whose length it runs for fifty miles. The southern twenty miles of its course across this desert plain is marked by a line of hills from two to three hundred feet high, which undoubtedly are the product of some movement in the remote past.

As we trace the line farther south, the topography indicates profound movements. A considerable area west of the Sunset oil district has had its drainage so disturbed that with the light rainfall in this region the streams have never broken across the ridges formed by the earth movements. The depressions are occupied by water during exceptionally wet years, but for the most of the time they are white alkali sinks.

Throughout nearly the whole course of the rift between the Cholame Valley and the valley of San Bernardino there may be seen the nearly obliterated effect of the Tejón earthquake of 1857. This is a low scarp or ridge, at the foot of a higher ridge of an earlier time. At one point in the Carrisa Plain four or five earthquake ridges appear within a width of a quarter of a mile.

Passing up a cañon fully three thousand feet deep, which is without doubt due to faulting in recent geological time, the rift crosses the San Emedio Mountains through the head of the cañon of the same name. Along this stretch it forms springs.

Passing over the divide at the head of the east fork of the San Emedio Cañon, we reach Cuddy Valley, where the sinking of a block of the earth, now a fertile well-watered valley, has led to the creation of an escarpment some miles long and from one hundred to three hundred feet high. Upon this escarpment old pine trees are standing.

Going to Tejón Pass we continue down a long cañon, whose northern walls have been greatly shattered by earthquakes. In this cañon we note the formation of great debris fans at the mouths of the gulches, the material having been derived from the broken rocks.

The rift makes a great turn to the east, across the San Emedio Mountains, but at Tejón Pass it turns again to a southeasterly direction, which it follows very regularly for two hundred miles.

For several miles southeast from Gorman Station, which lies just east of Tejón Pass, the old Los Angeles-Bakersfield stage road follows the valley of the rift.

In the earthquake of 1857 this road is reported to have been seriously broken up, and the surface at the present time seems to verify the statement. A few miles southeast of Gorman Station the rift passes out to Antelope Valley (the western arm of the Mohave Desert), skirting the lofty mountains which rise on the south.

By means of springs, cienegas, and long, narrow valleys, we trace the rift to a point one mile south of Palmdale, a station on the Southern Pacific Railroad. The rift passes through Lake Elizabeth, the largest body of water in this part of California. The lake itself is due to the interference of earth movements with the drainage of a broad valley.

Near Palmdale are interesting ridges and sinks. One of the latter is so large that it has been used as a reservoir to store water for irrigation. Looking from the ridge where the railroad crosses the rift, we can follow its course for many miles by a break in the slope of the desert wash.

Strikingly interesting features appear as we climb the northern flanks of the San Gabriel Range and pass out of the desert into the pine forests. At one spot a ridge has been split away from the mountain, making two streams where there was formerly only one. Upon this ridge there are large pine trees.

At an elevation of seven thousand feet we cross over the San Gabriel Range and descend a long, straight cañon, called Lone Pine, to the lower end of Cajón Pass. From here the rift marks the line between the lofty front of the San Bernardino Range and the gravel mesa leading down to the valley. Movements of the country upon either side of the rift have broken the mesa, producing ridges and scarps, some of which are over fifty feet high. Here, as throughout the course of the rift along the southern border of the Mohave Desert, the grinding movements have produced an impervious clay layer, which stops the waters coming down from the mountains and brings them to the surface, thus forming springs, cienegas, meadows, and lakes. The rift line is in fact one of the most important economic features of the desert region which it traverses, for nearly half its known length, producing surface water in a land where water is the most important consideration.

As we stand in the valley and look at the lofty San Bernardino Range, we can readily imagine its slow growth by repeated earthquake movements through the long ages of the past and how the region looked when the ancient and nearly worndown mountains of the present Mohave Desert extended unbroken across to the sea and far to the south.

East of the point where the Santa Ana River issues from the mountains a stream of considerable size has been compelled to form a new channel as a result of the uplift of the mesa gravels.

Beyond the Santa Ana River, in the direction of Potato Cañon, the features of the rift become more indistinct and in places are nearly obliterated. Crushing and breaking of the rocks on these steep mountain slopes has led to such rapid erosion that surface changes are comparatively rapid.

Southeast of Potato Cañon the rift line begins to turn more easterly and, instead of descending to and crossing the San Gorgonio Pass and then skirting the east base of San Jacinto Mountains, as it was at first thought to do, it was found to keep along the side of the mountains north of the pass and finally to disappear in the desert wash east of the Whitewater River. Where last seen, the course was due east, a direction which would carry it north of Palm Spring Station in the Conchilla Desert. While it is doubtful if the distinctive "earthquake topography" can be traced any farther, yet it is probable that a fault continues on still farther along the mountains lying north of the Salton Basin.

Although we think the earthquake of April, 1906, was severe, it was undoubtedly light when compared with many that have occurred along the same line. It will not be many years before the surface effects of the last earthquake will have generally disappeared. The effects of the Tejón earthquake are still visible although it took place fifty years ago. Imagination alone can picture the destructive effects of an earthquake which could form scarps 100 to 300 feet high.

The Temblor: A Personal Narration

By Mary Austin

There are some fortunes harder to bear once they are done with than while they are doing, and there are three things that I shall never be able to abide in quietness again—the smell of burning, the creaking of house-beams in the night, and the roar of a great city going past me in the street.

Ours was a quiet neighborhood in the best times; undisturbed except by the hawker's cry or the seldom whistling hum of the wire, and in the two days following April eighteenth, it became a little lane out of Destruction. The first thing I was aware of was being wakened sharply to see my bureau lunging solemnly at me across the width of the room. It got up first on one castor and then on another, like the table at a séance, and wagged its top portentously. It was an antique pattern, tall and marble-topped, and quite heavy enough to seem for the moment sufficient cause for all the uproar. Then I remember standing in the doorway to see the great barred leaves of the entrance on the second floor part quietly as under an unseen hand, and beyond them, in the morning grayness, the rose tree and the palms replacing one another, as in a moving picture, and suddenly an eruption of night-gowned figures crying out that it was only an earthquake, but I had already made this discovery for myself as I recall trying to explain. Nobody having suffered much in our immediate vicinity, we were left free to perceive that the very instant after the quake was tempered by the half-humorous, wholly American appreciation of a thoroughly good job. Half an hour after the temblor people sitting on their doorsteps, in bathrobes and kimonos, were admitting to each other with a half twist of laughter between tremblings that it was a really creditable shake.

San Francisco on the Night of April 18.

The appreciation of calamity widened slowly as water rays on a mantling pond. Mercifully the temblor came at an hour when families had not divided for the day, but live wires sagging across housetops were to outdo the damage of falling walls. Almost before the dust of ruined walls had ceased rising, smoke began to go up against the sun, which, by nine of the clock, showed bloodshot through it as the eye of Disaster.

Looking Southeast from Telegraph Hill During Fire

It is perfectly safe to believe anything any one tells you of personal adventure; the inventive faculty does not exist which could outdo the actuality; little things prick themselves on the attention as the index of the greater horror.

I remember distinctly that in the first considered interval after the temblor, I went about and took all the flowers out of the vases to save the water that was left; and that I went longer without washing my face than I ever expect to again.

I recall the red flare of a potted geranium undisturbed on a window ledge in a wall of which the brickwork dropped outward, while the roof had gone through the flooring; and the cross-section of a lodging house parted cleanly with all the little rooms unaltered, and the halls like burrows, as if it were the home of some superior sort of insect laid open to the microscope.

South of Market, in the district known as the Mission, there were cheap man-traps folded in like pasteboard, and from these, before the rip of the flames blotted out the sound, arose the thin, long scream of mortal agony.

Down on Market Street Wednesday morning, when the smoke from the burning blocks behind began to pour through the windows we saw an Italian woman kneeling on the street corner praying quietly. Her cheap belongings were scattered beside her on the ground and the crowd trampled them; a child lay on a heap of clothes and bedding beside her, covered and very quiet. The woman opened her eyes now and then, looked at the reddening smoke and addressed herself to prayer as one sure of the stroke of fate. It was not until several days later that it occurred to me why the baby lay so quiet, and why the woman prayed instead of flying.

Looking South from Lafayette Square, 4 p. m. April 18

Not far from there, a day-old bride waited while her husband went back to the ruined hotel for some papers he had left, and the cornice fell on him; then a man who had known him, but not that he was married, came by and carried away the body and shipped it out of the city, so that for four days the bride knew not what had become of him.

There was a young man who, seeing a broken and dismantled grocery, meant no more than to save some food, for already the certainty of famine was upon the city—and was shot for looting. Then his women came and carried the body away, mother and betrothed, and laid it on the grass until space could be found for burial. They drew a handkerchief over its face, and sat quietly beside it without bitterness or weeping. It was all like this, broken bits of human tragedy, curiously unrelated, inconsequential, disrupted by the temblor, impossible to this day to gather up and compose into a proper picture.

Van Ness Avenue, 10:30 p. m. April 19

Jefferson Square

The largeness of the event had the effect of reducing private sorrow to a mere pin prick and a point of time. Everybody tells you tales like this with more or less detail. It was reported that two blocks from us a man lay all day with a placard on his breast that he was shot for looting, and no one denied the aptness of the warning. The will of the people was toward authority, and everywhere the tread of soldiery brought a relieved sense of things orderly and secure. It was not as if the city had waited for martial law to be declared, but as if it precipitated itself into that state by instinct as its best refuge.

In the parks were the refugees huddled on the damp sod with insufficient bedding and less food and no water. They laughed. They had come out of their homes with scant possessions, often the least serviceable. They had lost business and clientage and tools, and they did not know if their friends had fared worse. Hot, stifling smoke billowed down upon them, cinders pattered like hail—and they laughed—not hysteria, but the laughter of unbroken courage.

That exodus to the park did not begin in our neighborhood until the second day; all the first day was spent in seeing such things as I relate, while confidently expecting the wind to blow the fire another way. Safe to say one-half the loss of household goods might have been averted, had not the residents been too sure of such exemption. It happened not infrequently that when a man had seen his women safe he went out to relief work and returning found smoking ashes—and the family had left no address. We were told of those who had dead in their households who took them up and fled with them to the likeliest place in the hope of burial, but before it had been accomplished were pushed forward by the flames. Yet to have taken part in that agonized race for the open was worth all it cost in goods.

Before the red night paled into murky dawn thousands of people were vomited out of the angry throat of the street far down toward Market. Even the smallest child carried something, or pushed it before him on a rocking chair, or dragged it behind him in a trunk, and the thing he carried was the index of the refugee's strongest bent. All the women saved their best hats and their babies, and, if there were no babies, some of them pushed pianos up the cement pavements.

All the faces were smutched and pallid, all the figures sloped steadily forward toward the cleared places. Behind them the expelling fire bent out over the lines of flight, the writhing smoke stooped and waved, a fine rain of cinders pattered and rustled over all the folks, and charred bits of the burning fled in the heated air and dropped among the goods. There was a strange, hot, sickish smell in the street if it had become the hollow slot of some fiery breathing snake. I came out and stood in the pale pinkish glow and saw a man I knew hurrying down toward the gutted district, the badge of a relief committee fluttering on his coat. "Bob," I said, "it looks like the day of judgment!" He cast back at me over his shoulder unveiled disgust at the inadequacy of my terms. "Aw!" he said, "it looks like hell!"

It was a well-bred community that poured itself out into Jefferson Square, where I lay with my friend's goods, and we were packed too close for most of the minor decencies, but nobody forgot his manners. "Beg pardon!" said a man hovering over me with a 200-pound trunk. "Not at all!" I answered making myself thin for him to step over. With an "Excuse me, madam!" another, fleeing from the too-heated border of the park to its packed center, deftly up-ended a roll of bedding, turned it across the woman who lay next to me—and the woman smiled.

Right here, if you had time for it, you gripped the large, essential spirit of the West, the ability to dramatize its own activity, and, while continuing in it, to stand off and be vastly entertained by it. In spite of individual heartsinkings, the San Franciscans during the week never lost the spirited sense of being audience to their own performance. Large figures of adventure moved through the murk of those days—Denman going out with his gun and holding up

"Nob Hill" After the Fire, Showing the Huntington, Crocker, and Flood Residences. Fairmont Hotel in Background.

express wagons with expensively saved goods, which were dumped out on sidewalks that food might be carried to unfed hundreds; Father Ramm cutting away the timbers of St. Mary's tower, while the red glow crept across the charred cross out of reach of the hose; and the humble sacrifices—the woman who shared her full breast with the child of another whose fountain had failed from weariness and fright—would that I had her name to hold in remembrance! She had stopped in the middle of a long residence hill and rested on a forsaken stoop, nourishing her child quietly, when the other woman came by panting, fainting and afraid, not of her class, nor her race, but the hungry baby yearned toward the uncovered breast—and they both of them understood that speech well enough.

In the Apartment House District

Tivoli Theatre

Everybody tells you tales like this, more, and better. All along the fire line of Van Ness Avenue, heroic episodes transpired like groups in a frieze against the writhing background of furnace-heated flame; and, for a pediment to the frieze, rows of houseless, possessionless people wrapped in a large, impersonal appreciation of the spectacle.

From Gough Street, looking down, we saw the great tide of fire roaring in the hollow toward Russian Hill; burning so steadily for all it burned so fast that it had the effect of immense deliberation; roaring on toward miles of uninhabited dwellings so lately emptied of life that they appeared consciously to await their immolation; beyond the line of roofs, the hill, standing up darkly against the glow of other incalculable fires, the uplift of flames from viewless intricacies of destruction, sparks belching furiously intermittent like the spray of bursting seas. Low down in front ran besmirched Lilliputians training inadequate hose and creating tiny explosions of a block or so of expensive dwellings by which the rest of us were ultimately saved; and high against the tip of flames where it ran out in broken sparks, the figure of the priest chopping steadily at the tower with the constrained small movement of a mechanical toy.

Observe that a moment since I said houseless people, not homeless; for it comes to this with the bulk of San Franciscans, that they discovered the place and the spirit to be home rather than the walls and the furnishings. No matter how the insurance totals foot up, what landmarks, what treasures of art are evanished , San Francisco, our San Francisco is all there yet. Fast as the tall banners of smoke rose up and the flames reddened them, rose up with it something impalpable, like an exhalation. We saw it breaking up in the movements of the refugees, heard it in the tones of their voices, felt it as they wrestled in the teeth of destruction. The sharp sentences by which men called to each other to note the behavior of brick and stone dwellings contained a hint of a warning already accepted for the new building before the old had crumbled. When the heat of conflagration outran the flames and reaching over wide avenues caught high gables and crosses of church steeples, men watching them smoke and blister and crackle into flame, said shortly, "No more wooden towers for San Francisco!" and saved their breath to run with the hose.

What distinguishes the personal experience of the destruction of the gray city from all like disasters of record, is the keen appreciation of the deathlessness of the spirit of living. For the greater part of this disaster—the irreclaimable loss of goods and houses, the violent deaths—was due chiefly to man-contrivances, to the sinking of made ground, to huddled buildings cheapened by greed, to insensate clinging to the outer shells of life; the strong tug of nature was always toward the renewal of it. Births near their time came on hurriedly; children were delivered in the streets or the midst of burnings, and none the worse for the absence of conventional circumstance; marriages were made amazingly, as the disorder of the social world threw all men back severely upon its primal institutions.

Fairmont Hotel After the Fire

After a great lapse of time, when earthquake stories had become matter for humorous reminiscence, burning blocks topics of daily news, and standing in the bread line a fixed habit—by the morning of the third day, to be exact—there arose a threat of peril greater than the thirst or famine, which all the world rose up swiftly to relieve.

Thousands of families had camped in parks not meant to be lived in, but to be looked at; lacking the most elementary means of sanitation. With the rising of the sun, a stench arose from these places and increased perceptibly; spreading with it like an exhalation, went the fear of pestilence. But this at least was a dread that every man could fight at his own camp, and the fight was the modern conviction of the relativity of sanitation to health. By mid-morning the condition of Jefferson Square was such that I should not have trusted myself to it for three hours more, but in three hours it was made safe by no more organized effort than came of the intelligent recognition of the peril. They cleaned the camp first, and organized committees of sanitation afterward.

There have been some unconsidered references of the earthquake disaster to the judgment of God; happily not much of it, but enough to make pertinent some conclusions that shaped themselves swiftly as the city fought and ran. Not to quarrel with the intelligence that reads God behind seismic disturbance, one must still note that the actual damage done by God to the city was small beside the possibilities for damage that reside in man-contrivances; for most man-made things do inherently carry the elements of their own destruction.

How much of all that happened of distress and inestimable loss could have been averted if men would live along the line of the Original Intention, with wide, clean breathing spaces and room for green growing things to push up between?

I have an indistinct impression that the calendar time spent in the city after the temblor was about ten days. I remember the night of rain, and seeing a grown man sitting on a curbstone the morning after, sobbing in the final break-down of bodily endurance. I remember too the sigh of the wind through windows of desolate walls, and the screech and clack of ruined cornices in the red noisy night, and the cheerful banging of pianos in the camps; the burials in trenches and the little, bluish, grave-long heaps of burning among the ruins of Chinatown, and the laughter that shook us as in the midst of the ashy desert we poured in dogged stream to the ferry, at a placard that in a half-burned building where activity had begun again, swung about in the wind and displayed this legend:

DON'T TALK EARTHQUAKE **TALK BUSINESS**

All these things seem to have occurred within a short space of days, but when I came out at last at Berkeley—too blossomy, too full-leafed, too radiant—by this token I knew that a great hiatus had taken place. It had been long enough to forget that the smell of sun-steeped roses could be sweet.